Ultraviolet

and Visible Spectroscopy

ANALYTICAL CHEMISTRY BY OPEN LEARNING

ACOL (Analytical Chemistry by Open Learning) is a well established series which comprises 33 open learning books and 9 computer based training packages. This open learning material covers all of the important techniques and fundamental principles of analytical chemistry.

Books

Samples and Standards
Sample Pretreatment
Classical Methods Vols I and II
Measurement, Statistics and Computation
Using Literature
Instrumentation
Chromatographic Separations
Gas Chromatography
High Performance Liquid Chromatography
Electrophoresis
Thin Layer Chromatography
Ultraviolet and Visible Spectroscopy
Fluorescence and Phosphorescence
Atomic Absorption and Emission
    Spectroscopy
Nuclear Magnetic Resonance
    Spectroscopy
X-Ray Methods
Mass Spectrometry

Scanning Electron Microscopy and
    Microanalysis
Principles of Electroanalytical Methods
Potentiometry and Ion Selective Electrodes
Polarography and Voltammetric Methods
Radiochemical Methods
Clinical Specimens
Diagnostic Enzymology
Quantitative Bioassay
Assessment and Control of Biochemical
    Methods
Thermal Methods
Microprocessor Applications
Chemometrics: Experimental Design
Environmental Analysis
Quality in the Analytical Chemistry
    Laboratory
Modern Infrared Spectroscopy

Software

Atomic Absorption Spectroscopy
High Performance Liquid Chromatography
Polarography
Radiochemistry
Gas Chromatography
Fluorescence
Quantitative IR and UV
Chromatography
Measure Your Own Quality

Series Editor: David J. Ando

Further information: ACOL–BIOTOL Office
                       The University of Greenwich
                       Unit 42, Butterly Avenue
                       Dartford Trade Park
                       Dartford
                       DA1 1JG

Ultraviolet and Visible Spectroscopy

Analytical Chemistry by Open Learning
Second Edition

Author:
MICHAEL J. K. THOMAS
University of Greenwich

Editor:
DAVID J. ANDO
University of Greenwich

Authors of First Edition:
RONALD C. DENNEY
University of Greenwich

ROY SINCLAIR
University of Paisley

Published on behalf of ACOL (University of Greenwich)
by
JOHN WILEY & SONS
Chichester • New York • Brisbane • Toronto • Singapore

Other Wiley Editorial Offices

John Wiley & Sons, Inc., 605 Third Avenue,
New York, NY 10158-0012, USA

Jacaranda Wiley Ltd, 33 Park Road, Milton,
Queensland 4064, Australia

John Wiley & Sons (Canada) Ltd, 22 Worcester Road,
Rexdale, Ontario M9W 1L1, Canada

John Wiley & Sons (Asia) Pte Ltd, 2 Clementi Loop #02-01,
Jin Xing Distripark, Singapore 129809

Library of Congress Cataloging-in-Publication Data

Thomas, Michael J.K.
    Ultraviolet and visible spectroscopy   /   author, Michael J.K. Thomas
    ; editor, David J. Ando.
        p.   cm.—(Analytical Chemistry by Open Learning)
    Includes bibliographical references and index.
    ISBN 0-471-96742-4 (cloth   :   alk. paper).—ISBN 0-471-96743-2
    (pbk. :   alk. paper)
        1. Spectrum analysis—Programmed instruction.   2. Chemistry,
    Analytic—Programmed instruction.   I. Ando, D.J. (David J.)
    II. Title.   III. Series: Analytical Chemistry by Open Learning
    (Series)
    QD95.T46   1996   96-11993
    543'.08584—dc20   CIP

British Library Cataloguing in Publication Data

A catalogue record for this book is available from the British Library

ISBN 0 471 96742 4 (cloth)
ISBN 0 471 96743 2 (paper)

Typeset in 11/13pt Times by Mackreth Media Services, Hemel Hempstead, Herts
Printed and bound in Great Britain by Biddles Ltd, Guildford, Surrey
This book is printed on acid-free paper responsibly manufactured from sustainable forestation,
for which at least two trees are planted for each one used for paper production.

THE UNIVERSITY OF GREENWICH
ACOL PROJECT

This series of easy-to-read books has been written by some of the foremost lecturers in Analytical Chemistry in the United Kingdom. These books are designed for training, continuing education and updating of all technical staff concerned with Analytical Chemistry.

These books are for those interested in Analytical Chemistry and instrumental techniques who wish to study in a more flexible way than traditional institute attendance, or to augment such attendance.

ACOL also supply a range of training packages which contain computer software together with the relevant ACOL book(s). The software teaches competence in the laboratory by providing experience of decision making in such an environment, often based on the simulation of instrumental output, while the books cover the requisite underpinning knowledge.

The Royal Society of Chemistry uses ACOL material to run regular series of courses based on distance learning and regular workshops.

Further information on all ACOL materials and courses may be obtained from:

ACOL–BIOTOL Office, University of Greenwich, Unit 42, Butterly Avenue, Dartford Trade Park, Dartford, DA1 1JG. Tel: 0181-331-9600, Fax: 0181-331-9672.

How to Use an Open Learning Book

Open Learning books are designed as a convenient and flexible way of studying for people who, for a variety of reasons, cannot use conventional education courses. You will learn from this book the principles of one subject in Analytical Chemistry, but only by putting this knowledge into practice, under professional supervision, will you gain a full understanding of the analytical techniques described.

To achieve the full benefit from an open learning text you need to carefully plan your place and time of study.

- Find the most suitable place to study where you can work without disturbance.

- If you have a tutor supervising your study discuss with this person the date by which you should have completed the text.

- Some people study perfectly well in irregular bursts; however, most students find that setting aside a certain number of hours each day is the most satisfactory method. It is for you to decide which pattern of study suits you best.

- If you decide to study for several hours at once, take short breaks of five or ten minutes every half hour or so. You will find that this method maintains a higher overall level of concentration.

Before you begin a detailed reading of this book, familiarise yourself with the general layout of the material. Have a look at the course contents list at the front of the book and flip through the pages to get a general impression of the way the subject is dealt with. You will find that there is space on the pages to make comments alongside the text as you study — your own notes for highlighting points that you feel are particularly important. Indicate in the margin the points you would like to discuss further with a tutor or fellow student. When you come to revise, these personal study notes will be very useful.

Π   When you find a paragraph in the text marked with a symbol such as is shown here, this is where you get involved. At this point you are directed to do certain things, e.g. draw graphs, answer questions, perform calculations, etc. Do make an attempt at these activities. If necessary, cover the succeeding response with a piece of paper until you are ready to read on. This is an opportunity for you to learn by participating in the subject, and although the text continues by discussing your response, there is no better way to learn than by working things out for yourself.

We have introduced self-assessment questions (SAQs) at appropriate places in the text. These SAQs provide you with a way of finding out if you understand what you have just been studying. There is space on the page for your answer and for any comments you want to add after reading the author's response. You will find the author's response to each SAQ at the end of the book. Compare what you have written with the response provided and read the discussion and advice.

At intervals in the text you will find a Summary and a list of Objectives. The Summary will emphasise the important points covered by the material you have just read, while the Objectives will give you a checklist of tasks you should then be able to achieve.

You can revise the book, perhaps for a formal examination, by re-reading the Summary and the Objectives, and by working through some of the SAQs. This should quickly alert you to areas of the text that need further study.

At the end of this book you will find, for reference, lists of commonly used scientific symbols and values, units of measurement, and also a periodic table.

Contents

Study Guide

The first Ultraviolet and Visible Spectroscopy Unit in the ACOL series was written by Ron Denney and Roy Sinclair. The revision for the Second Edition of this Unit has been carried out by Mike Thomas, who is currently lecturing at the University of Greenwich, and who has added to and reorganised the earlier material in order to bring it completely up to date.

The overall aims of this present text have not changed from the First Edition, namely to provide you with a working knowledge of modern ultraviolet/visible spectroscopy and the ability to apply the technique to a range of analytical problems.

Little prior knowledge is assumed, although it is likely that you will have an understanding of chemistry which is equivalent to that of a student who has studied HNC or HND in BTEC chemistry, or GNVQ Level 4. The ability to rearrange and solve simple algebraic expressions is the only mathematics that is required.

The Unit is designed to be an introduction to the theory and application of ultraviolet/visible spectroscopy in mainly quantitative analysis and you may find that there are areas in which you are particularly interested and for which a more detailed account is necessary. Suitable books that you could use for this purpose are listed in the Bibliography. You may find that some areas are not particularly relevant to your needs, for example the final chapter, and therefore you may wish to treat these topics as optional.

Chemistry in general, and analytical chemistry in particular, is a practical subject and so the best way in which to test your understanding of this Unit is to carry out some colorimetric or ultraviolet analyses of your own. It is only in this way that you will appreciate the many pitfalls which may occur when carrying out an analysis using ultraviolet/visible spectroscopy.

Supporting Practical Work

The aims of the practical work are as follows:

(a) To provide a basic experience of using a spectrometer for recording ultraviolet/visible spectra.

(b) To give experience in the use of ultraviolet/visible spectroscopy in quantitative analysis, with particular emphasis on good analytical practice.

In order to achieve these aims you could repeat some of the exercises described in the text, although the determination of glucose in a foodstuff does require a specialist test kit. Other similar experiments could be equally appropriate in realising the aims for this Unit.

Bibliography

ANALYTICAL CHEMISTRY TEXTBOOKS

All books in this subject area contain major chapters which deal with ultraviolet/visible spectroscopy. The following are used extensively:

(a) G. H. Jeffery, J. Bassett, J. Mendham and R. C. Denney, *Vogel's Textbook of Quantitative Chemical Analysis*, 5th Edn, Longmans, 1989.

(b) F. W. Fifield and D. Kealey, *Principles and Practice of Analytical Chemistry*, 2nd Edn, International Textbook Company Ltd, 1983.

(c) H. H. Willard, L. L. Merritt, J. A. Dean and F. A. Settle, *Instrumental Methods of Analysis*, 6th Edn, Wadsworth Publishing Co., 1981.

(d) D. C. Harris, *Quantitative Chemical Analysis*, W. H. Freeman and Co., 1982.

TEXTBOOKS ON ABSORPTION SPECTROSCOPY

These books deal with the theory and practice of spectroscopy, with special chapters on ultraviolet/visible spectra treated in depth.

(a) J. M. Hollas, *Modern Spectroscopy*, Wiley, 1987.

(b) C. N. Banwell and E. M. McCash, *Fundamentals of Molecular Spectroscopy*, 4th Edn, McGraw-Hill, 1994.

Acknowledgements

Extracts in Section 3.1.1 on the determination of iron in raw and potable waters are taken from *Iron in Raw and Potable Waters by Spectrophotometry* (1977), published by HMSO. Crown copyright is reproduced with the permission of the Controller of HMSO.

The method given in Section 3.1.2 for the determination of sucrose and D-glucose in foodstuffs is reproduced by the permission of Boehringer Mannheim GmbH.

Extracts in Section 3.1.3 on the determination of trace amounts of iron in pure reagent chemicals are taken from the *Analyst*, **101**, 974 (1976) and are reproduced by permission of the Royal Society of Chemistry.

Figure 3.3c is redrawn from R. A. Morton and A. L. Stubbs, *Analyst*, **71**, 348 (1946) and is reproduced by permission of the Royal Society of Chemistry.

Figures 4.1a and 4.1b are redrawn from a Varian Applications leaflet and are reproduced with permission.

1. Introduction

The foundations of quantitative chemical analysis can be traced back to the development of titrimetric analysis in which titration end-points depended on the change in colour of the species being analysed or of that of a specially added chemical indicator. These colour changes arise due to molecular and structural changes in the substances being examined, leading to changes in the ability to absorb light in the visible region of the electromagnetic spectrum. In various ways absorption spectroscopy in the visible region has long been an important tool to the analyst. Many important and sensitive colour tests have been developed for the detection and determination of a wide range of chemical species, both inorganic and organic in nature, and were used long before the development of visible and ultraviolet spectrometers.

Today, ultraviolet (UV)/visible spectroscopy is applied to many thousands of determinations which have been developed over the years. It has proved particularly useful in biochemical analysis, and is of vital importance in the clinical laboratory attached to most modern hospitals where various components of blood and/or urine samples, in particular, are determined and monitored on a twenty-four-hour basis. It plays a part in environmental studies of pollutants, in forensic-science work on drugs, and in maintaining the quality of the food that we eat. In all of these areas UV/visible spectroscopy is an essential tool in the identification and quantification of a very broad range of chemical and biological substances. The equipment for these purposes ranges from very simple colour comparators through to large computer-controlled automatic scanning instruments, which cover the whole of the UV/visible region of the electromagnetic spectrum. In all instances, however, these studies involve measurement of radiation intensity at the spectral wavelengths which are characteristic of the substances under investigation.

1.1   COLOUR TESTS AND CHEMICAL ANALYSIS

One of the earliest tests which you may have encountered in your study of chemistry is the change in colour of anhydrous copper(II) sulfate crystals from white to blue when water is added, or the change from red to blue of litmus paper when it is dipped in an alkaline solution. These, and similar observations of colour and colour changes, would probably have been your first experiences of the principles of absorption spectroscopy as applied to the study of chemical systems. A material will appear coloured if it shows selective absorption of radiation within the visible region of the electromagnetic spectrum and any change in that absorption will be associated with a change in colour. With modern instrumentation, however, it is no longer necessary for the chemical species to be coloured for its selective absorption to be studied.

The observation of colour or a colour change has led to significant discoveries in the field of chemistry and the development of new materials and dyestuffs.

1.1.1   Colour Tests and Qualitative Chemical Analysis

Simple colour tests have often proved to be useful preliminary or confirmatory evidence for the presence of a particular chemical species. You will, therefore, already have encountered a number of characteristic colour tests. Some possible examples are listed below:

(a) The use of litmus or indicator paper to test acidity/alkalinity.

(b) The yellow colour imparted to a gas flame when testing for sodium ions.

(c) The deep blue colour produced when testing for iodine by using starch solution.

These tests are probably familiar to you. Six further well known tests are listed below.

∏ Can you identify which of these tests involve an observation of colour or colour change? Try to write down the chemical equation involved in each test.

(a) The test for halide ions using silver nitrate solution.

(b) The test for Mn^{2+} ions in aqueous solution with hydrogen sulfide gas.

(c) The test for Fe^{3+} ions with aqueous potassium thiocyanate solution.

(d) The test for aldehydes with Fehling's solution.

(e) The test for ketones with 2,4-dinitrophenylhydrazine.

(f) The Lassaigne test for nitrogen in organic compounds.

The correct response is that the appearance of a characteristic colour is the main observation in tests (c), (d), and (f). However, with tests (a), (b), and (e), the main observation is the formation of a precipitate, although the colour of the precipitate may also be of some significance, as indicated below. The nature of the chemical reactions involved are indicated in the following notes.

Test (a)

The presence of a halide (X^-) is indicated by a precipitate of AgX, which for chloride is white, for bromide is a very pale yellow colour, and for iodide is distinctly yellow. The colour of the precipitate is, therefore, a useful part of the test.

$$X^- + AgNO_3 \longrightarrow AgX + NO_3^-$$

Test (b)

A number of metal sulfides, including copper, cadmium and lead, are readily precipitated in acid solution. Manganese ions, however, are precipitated under alkaline conditions to give a pale creamy pink precipitate of manganese sulfide. Again, the colour of the precipitate may aid in the identification.

$$Mn^{2+} + H_2S \longrightarrow MnS + 2H^+$$

Test (c)

Aqueous solutions of iron(III) give a characteristic blood red colour with potassium thiocyanate. This is a very sensitive test for low concentrations of iron in the 3+ state. Although the reaction shown is an equilibrium the equilibrium, constant is sufficiently high that this colour reaction can be used for the quantitative determination of iron(III).

$$[Fe(H_2O)_6]^{3+} + SCN^- \rightleftharpoons [Fe(H_2O)_5SCN]^{2+} + H_2O$$

✗ *Test (d)*

Fehling's solution is an alkaline solution of copper(II) complexed with tartrate ion, and is initially blue in colour. If this solution is warmed in the presence of an aldehyde, the copper(II) is reduced and brick-red copper(I) oxide is precipitated.

$$RCHO + Cu^{2+} \longrightarrow Cu_2O + RCOOH$$

✳ *Test (e)*

Both aldehydes and ketones react, through the carbonyl group (C=O), with 2,4-dinitrophenylhydrazine to give yellow or red precipitates of the corresponding hydrazone, which possess characteristic melting points.

Test (f)

$$CN^- + Fe^{2+}/Fe^{3+} \longrightarrow Fe_3[Fe(CN)_6]_2$$

The Lassaigne test for nitrogen involves the formation of sodium cyanide, which when boiled with iron(II) sulfate gives a solution of sodium hexacyanoferrate(II). If this solution is acidified and a few drops of iron(III) solution added the characteristic colour of Prussian Blue confirms the presence of nitrogen in the organic compound.

Before looking at the nature of the processes which lead to absorption in the ultraviolet or visible part of the electromagnetic spectrum, it is of interest to consider some examples from the early history of chemical analysis in which the observation of colour played an important role.

1.1.2 Some Historical Aspects of Colour Tests

Many of the chemical reactions discovered in the late 18th century were utilised in the development of methods of analysis in the early 19th century, and most of these involved either characteristic colour reactions or precipitations of the types considered above.

For example, the preparation of Prussian Blue was discovered by a colour manufacturer in Berlin in 1704 when he boiled iron(II) sulfate with a solution of potassium carbonate contaminated with cyanide. Later, a French dyer digested Prussian Blue with potassium hydroxide and obtained 'yellow prussiate of potash' or, as we would now call it, potassium hexacyanoferrate(II), $K_4[Fe(CN)_6]$. As there was considerable interest in mineral analysis at that time this compound quickly became the preferred reagent for the identification of iron(III). W. T. Brande at the Royal Institution in London extended this work and showed that the reagent was able to give characteristic coloured precipitates with 16 different metal ions, including titanium and uranium.

It was not until 1822, however, that the corresponding reagent, 'red prussiate of potash', i.e. potassium hexacyanoferrate(III), $K_3[Fe(CN)_6]$, was prepared by passing chlorine gas through a solution of the yellow prussiate. This new reagent was quickly adopted for testing for iron(II), since it also gives a deep blue precipitate, which is called Turnbull's Blue. It was not until over a hundred years later that it was shown that Prussian Blue and Turnbull's Blue are indeed the same compound.

Scheele was another researcher who encountered numerous colour changes of compounds during his work which lead to the discovery of chlorine in 1774. He fused together manganese dioxide and potassium nitrate and extracted the solid mass with water to obtain a green solution, which turned a purplish-red colour on dilution. In the process he converted potassium manganate(VI), K_2MnO_4, to potassium manganate(VII), $KMnO_4$. It was found that in acid solution the purple salt was easily reduced to the almost colourless Mn^{2+} species, a reaction which was soon applied to titrimetric analysis of, for example, iron(II), hydrogen peroxide and ethandioic (oxalic) acid.

These are just two examples of the importance of colour changes in the history of chemical analysis; and many others also exist. Since the ability to detect colour changes and to observe colour has been, and still is, of great importance in analysis, it is relevant to consider the way in which light absorption produces the various colours and how we are able to interpret these absorptions as colours of different hues through our eyes.

1.1.3 The Electromagnetic Spectrum

Electromagnetic radiation is a type of energy that is transmitted through space at enormous velocities. It takes many forms; light is the most easily recognised, but it also includes X-rays, and ultraviolet, radio and microwave radiation. In order to characterise many of the properties of electromagnetic radiation it is necessary to ascribe the wave nature to its propagation. The various types of radiation can then be defined in terms of their wavelengths or frequencies. The wavelength, λ, is defined as the linear distance between successive maxima or minima of a wave, and the frequency, ν, as the number of waves per second. The unit of frequency is usually the hertz (Hz), which is equivalent to one cycle per second. Multiplication of the wavelength (in metres per cycle) by the frequency (in Hz) gives the velocity of the radiation (in metres per second) as follows:

$$c = \nu\lambda \tag{1.1}$$

where c is the velocity of light in a vacuum and has the numerical value $3 \times 10^8\,\mathrm{m\,s^{-1}}$.

It is therefore possible to calculate the wavelength of radiation, given its frequency and vice versa, so that:

$$\text{wavelength } (\lambda) = \frac{\text{velocity of light } (c)}{\text{frequency } (\nu)} \qquad (1.2)$$

∏ Use the above expression and the given value for the velocity of light to calculate the frequency of green light having a wavelength of 500 nanometres (nm).

As a first step, the wavelength must be converted into 500×10^{-9} m. This ensures that we use the same units throughout the calculation. Substituting into equation (1.2) and rearranging, you should find that the frequency is 6×10^{14} Hz.

Another expression which is very important in all forms of spectroscopy relates the frequency of the electromagnetic radiation to its energy. The energy (in joules, J) is proportional to the frequency of the radiation as follows:

$$E = h\nu \qquad (1.3)$$

where h is a constant, known as the Planck constant. It has the numerical value of 6.624×10^{-34} J s.

∏ Use the value for the frequency obtained above to calculate the energy of the green light.

The frequency for the green light that we calculated above was 6×10^{14} Hz. If we substitute this value into equation (1.3) we get an energy of 3.974×10^{-19} J. If we multiply this value by the Avogadro constant $(6.023 \times 10^{23} \text{ mol}^{-1})$ the energy will then be in J mol^{-1}. This value is 239 378 J mol^{-1} or 239.378 kJ mol^{-1}.

The electromagnetic spectrum covers an immense range of wavelengths or energies. The various regions of the spectrum are shown in Figure 1.1a. The visible region of the electromagnetic spectrum is defined in terms of the wavelength range to which the human eye responds. The usual range of wavelengths quoted is from 380 nm at the blue/violet end to 750 nm at the long wavelength, i.e. red end of the spectrum. Some sources, however, quote wavelengths down

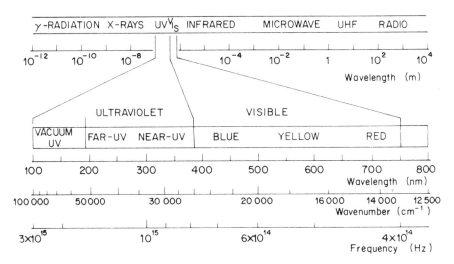

Figure 1.1a   The range of electromagnetic radiation

to 360 nm as the short-wavelength limit and as high as 780 nm as the long-wavelength limit. The ultraviolet region extends from about 380 to 100 nm. Below 200 nm, however, oxygen in the air makes it difficult to record the absorption of the radiation by the chemical species of interest. For this reason the 'accessible ultraviolet' region is usually taken to be from 200 to 400 nm. The lower limit is determined mainly by instrumental factors, such as the lack of detector sensitivity, and the reduced transmittance of radiation by the optical components, as well as the increased absorption of radiation by common solvents at low wavelengths.

You should note that as the wavelength of the electromagnetic radiation decreases its energy increases, and so red light is at the low-energy end of the visible spectrum and blue light is at the high-energy end.

In this Unit we are only concerned with the visible and the accessible part of the ultraviolet region covering the total wavelength range from about 200 to 800 nm.

When investigating the wavelengths of UV/visible radiation at which a particular sample absorbs, it is useful to produce a spectrum, which is

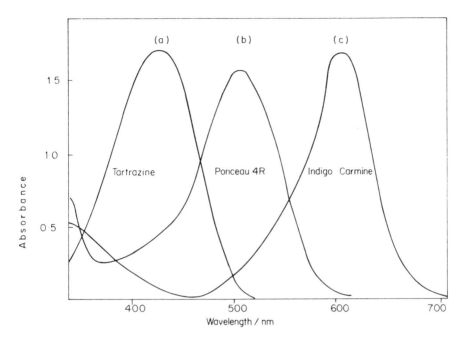

Figure 1.1b Visible absorption spectra

simply a plot of the amount of radiation absorbed versus the wavelength of the radiation. We will be meeting many spectra in this Unit, and the first of these are given in Figure 1.1b. Note that none of the spectra cover the entire region (200–800 nm) in which we are interested. Most of the spectra you will come across will only cover a selected part of the UV/visible region, in order to reflect the particular wavelength range over which the particular sample absorbs.

SAQ 1.1a

> Wavenumber ($\bar{\nu}$) values in cm^{-1} units are calculated by taking the reciprocal of the wavelength (λ) values, and multiplying these by an appropriate factor to allow for the conversion of units.
>
> (i) What is the relationship between wavelength values in nm and wavenumbers in cm^{-1} units?
>
> $\longrightarrow$

SAQ 1.1a
(Contd)

(ii) Similarly, what is the relationship between wavenumber values in cm^{-1} units and frequency values in hertz (or s^{-1})?

Use these two relationships to calculate the wavenumber and the frequency of yellow radiation of wavelength 575 nm. Check your values against the wavenumber and frequency scales of Figure 1.1a.

1.1.4 Light Absorption and Colour

Figure 1.1b shows typical absorption spectra of a yellow, a red and a blue dye, as follows:

(a) tartrazine or food yellow 4;

(b) Ponceau 4R or food red 7;

(c) indigo carmine or food blue 1.

These examples have been chosen to illustrate that the absorption band moves to a longer wavelength as the principal colour changes in the order yellow, orange, red, purple, and blue. The colour observed is determined by the spectral distribution of the transmitted radiation, i.e. by the combined colour produced from the mixture of wavelengths that are *not* absorbed.

In order to define the absorption characteristics of a particular compound we could quote both the wavelength of the maximum absorption, λ_{max}, and the range of wavelengths over which strong absorption occurs. In practice, it is usual to quote the range of wavelengths at half the absorption maximum, i.e. the width at half peak-height.

∏ Can you obtain these two quantities (to the nearest 10 nm) for the three dyes shown in Figure 1.1b?

Compare your figures with those given below. Did you appreciate the meaning of the absorption range at half peak-height?

Colour of dye (name)	Absorption wavelength maximum (nm)	Wavelength range at half peak-height (nm)
Yellow (tartrazine)	430	380–480
Red (Ponceau 4R)	510	450–550
Blue (indigo carmine)	610	570–650

Therefore, the yellow dye, tartrazine, absorbs strongly, mainly over the violet/blue region, transmitting a mixture of green and red wavelengths which when seen together give the yellow colour which is observed. In contrast, the bluish-red dye Ponceau 4R, absorbs in the blue to green region, while transmitting freely at the red and violet wavelengths, with this mixture again producing the observed colour. The general relationship between absorption position, colour of absorbed light and resulting colour observed for the transmitted light, is shown in Figure 1.1c.

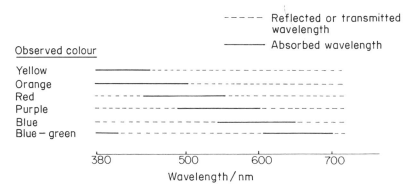

Figure 1.1c   Absorption band and colour relationships

Note that the wavelength ranges that are listed define the positions at which maximum absorption is observed. You may find that these ranges and limits may differ slightly from those quoted in other textbooks, although the trends will be similar.

You may have noticed that the wavelength range for a true green colour was omitted from the data in Figure 1.1c. This is because in order to see a green colour it is necessary to have a material which absorbs at both the blue and red ends of the visible spectrum and transmits only in the middle or green region. This requires the species in solution to show two absorption bands in the visible region.

The above explanation is a simplification of the relationship between light absorption and observed colour. In particular, the absorption spectra for the three dyes discussed here have half-bandwidths of about 100 nm and are classed as narrow absorption bands, and therefore give rise to quite vivid colours. Most coloured materials, particularly solids, show much broader absorption bands than this. In addition, the fact that the human eye is reputed to be able to distinguish over 1 million different shades of colour illustrates the limitations of relationships such as those shown in Figure 1.1c.

To test your understanding of the relationships involving the position of absorption and colour you should attempt the self-assessment question which follows. It involves distinguishing the absorption curves of some well known reagents found in most laboratories and which can be readily identified by the colour of their solutions. You will need to consider the effects of concentration on the amount of radiation that is absorbed.

SAQ 1.1b

Using your knowledge of the colours of the common reagent solutions listed below, identify the solutions corresponding to the spectra A to E in Figure 1.1d.

Solution	Reagent (concentration)
1	aqueous copper sulfate solution (0.4 mol dm^{-3});
2	aqueous copper sulfate solution (0.04 mol dm^{-3});
3	aqueous potassium dichromate solution (0.02 mol dm^{-3});
4	potassium dichromate (100 mg dm^{-3}) in dilute sulfuric acid (0.005 mol dm^{-3});
5	aqueous potassium permanganate solution (5×10^{-4} mol dm^{-3}).

Comment, if possible, on the relative intensities of absorption (relative absorptivities) of the three compounds.

→

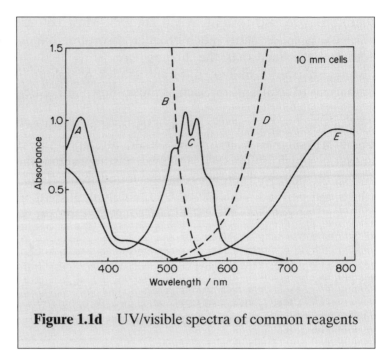

Figure 1.1d   UV/visible spectra of common reagents

SAQ 1.1b

For a colour test to have high sensitivity it is important for the coloured material produced in the test to absorb strongly in the visible region. Thus, as indicated in the response to SAQ 1.1b, the dichromate and copper(II) sulfate solutions are weakly coloured because the main absorptions lie outside the visible region. In order to produce a sensitive colour test for copper(II) and dichromate in aqueous solution, we need to use reagents which will produce species which will absorb strongly in the visible region of the spectrum.

As an example, a solution of 0.01 mol dm^{-3} $CuSO_4$ has a pale green/blue colour, but when a few drops of concentrated ammonia solution are added, a white precipitate is initially formed which then redissolves in excess ammonia to produce a solution having an intense blue colour. The changes in the absorption spectrum are shown in Figure 1.1e.

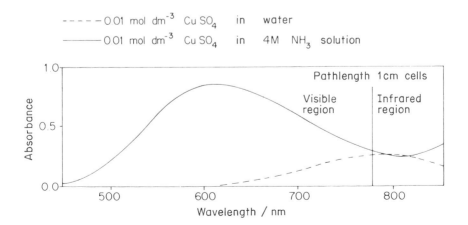

Figure 1.1e   Change in the absorption spectrum of copper sulfate solution upon addition of ammonia

In this case, the development of the stronger blue colour arises mainly because the absorption moves into the visible region, although there is also an increase in the intensity of the absorption.

Of course, if we are going to use an instrumental method for detection, rather than relying on our eyes, the requirement that the sample absorbs in the visible region of the spectrum is unimportant.

Some questions that might arise from the discussion of the copper(II) sulfate solution are as follows:

(a) Is the formation of the blue copper colour characteristic of only copper(II) sulfate solution or does it occur for all copper(II) species?

(b) What causes the increase in intensity of absorption when ammonia is added?

(c) Can the test be made quantitative?

(d) Why are both solutions blue, even though the absorption curves are significantly different?

The answers to the first three questions essentially form the subject of Chapter 2 of this Unit.

1.2   BEER'S LAW AND CALIBRATION DATA

We have already seen in SAQ 1.1b that the amount of radiation absorbed by a chemical species in solution is affected by the concentration of the solution. The measurement of the absorption of ultraviolet and visible radiation by species in solution provides one of the most widely used methods of quantitative analysis available in the analytical laboratory. The bases of such measurements are as follows:

(a) The generation of a suitably absorbing species in solution in amounts which are quantitatively related to the amount of analyte to be determined.

(b) The selection of a suitable wavelength to enable accurate measurements to be made of the absorbing species.

(c) The determination of the ratio of the intensity of the radiation on passing through a given thickness of the absorbing solution (usually held in a sample cell, known as a cuvette, of known pathlength) compared with the initial intensity of the same radiation beam.

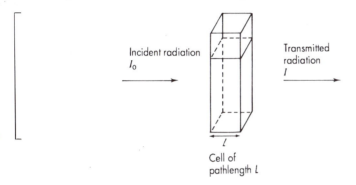

Incident radiation I_0

Transmitted radiation I

Cell of pathlength l

When light is absorbed by the solution the intensity of the incident beam, I_0, is reduced to I. We can define a term called the *transmittance*, T, as the ratio of the transmitted intensity to the incident intensity, as follows:

$$T = I/I_0 \tag{1.4}$$

Many instruments, particularly infrared spectrometers, often measure percentage transmittance, because they are calibrated with the very convenient scale of 0 to 100:

$$\% T = 100T = 100I/I_0 \tag{1.5}$$

A more useful measure for the amount of radiation absorbed by a solution is the *absorbance*, A. This is defined as follows:

$$A = \log (I_0/I) = \log (1/T) \tag{1.6}$$

and
$$A = \log (100/\% T) \tag{1.7}$$

The reason why we need to use a logarithmic term can be easily illustrated by considering the change of transmittance with the pathlength.

Suppose that we have a solution of potassium manganate(VII) which appears purple because it has a maximum absorption of radiation in the green region of the spectrum at 530 nm, and we look at the way in which the intensity of green light changes as it passes through glass cells of differing pathlength containing the solution. If after passing through 0.2 cm the light intensity has decreased by 50% of its initial value, then Lambert's law of light absorption, first proposed in 1760, would tell us that with a cell of 0.4 cm pathlength the intensity would have dropped by another 50%, or 25% of the original value. This would continue as shown in Figure 1.2a, until by a pathlength of 1 cm the overall transmittance would be 3.125% of the original intensity. This plot of the transmittance against the pathlength is an exponential curve. To obtain a linear plot we need to use a logarithmic function of the transmittance, which is the absorbance.

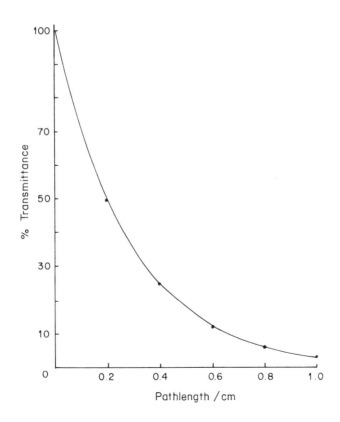

Figure 1.2a   Change of transmittance with pathlength

The table below shows the previous transmittance values converted into absorbances, from which it can be seen from Figure 1.2b that the plot of absorbance against pathlength is linear.

Pathlength (cm)	% Transmittance	$T = I/I_0$	$1/T$	$A = \log(1/T)$
0.0	100	1	1	0.000
0.2	50.0	0.5	2	0.301
0.4	25.0	0.25	4	0.602
0.6	12.5	0.125	8	0.903
0.8	6.25	0.0625	16	1.204
1.0	3.125	0.031 25	32	1.505

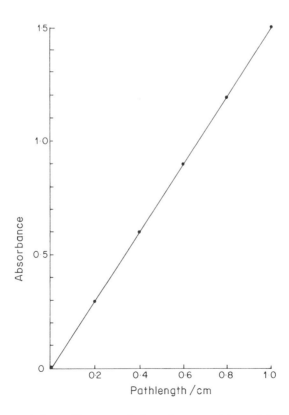

Figure 1.2b   Change of absorbance with pathlength

The above treatment is for a fixed concentration of the potassium manganate(VII) solution, but another law, i.e. Beer's law, which was proposed in 1852, also tells us that the absorbance depends on the total amount of the absorbing species in the radiation path through the cell, and this means that the absorption is affected by both the concentration (c) and the pathlength (l).

The combined Beer–Lambert law, now usually referred to as Beer's law, can be written in the following simple form:

$$A = acl \qquad (1.8)$$

where a is a proportionality constant, which is known as the *absorptivity*. Absorptivity is a constant for a particular species in a particular solvent at a particular wavelength, but it may have different numerical values depending on the units of concentration employed. In the specific case where the concentration of the compound in question is expressed in mol dm^{-3} it is usual to give the absorptivity a new symbol, ϵ (epsilon), which represents the *molar absorptivity* (also known in older textbooks as the molar absorption coefficient). Beer's law can then be rewritten as follows:

$$\log (I_0/I) = A = \epsilon cl \qquad (1.9)$$

In order to apply this simple law to the determinations of an analyte species of unknown concentrations in solution, it is necessary to first construct a calibration graph of absorbance versus concentration by using standard solutions of known concentrations of the analyte. The absorbance of the unknown can then be measured and its concentration read off from the calibration graph.

As we will see later, Beer's law applies to a solution containing more than one absorbing species in a solution, provided that there is no interaction between the various components. For a solution containing more than one absorbing species, the following applies:

$$A_{total} = A_1 + A_2 + \ldots + A_n$$

$$= \epsilon_1 l c_1 + \epsilon_2 l c_2 + \ldots + \epsilon_n l c_n \qquad (1.10)$$

where the subscripts refer to the absorbing components, $1, 2, \ldots, n$.

Strictly, Beer's law applies only to monochromatic radiation (i.e. radiation consisting of only one wavelength) and depends on the absorbing system (the solution, in this case) being homogeneous. In practice, we find that the law can be applied even when using radiation having a spread of wavelengths (polychromatic radiation).

SAQ 1.2a

Carry out the following calculations:

(i) Obtain a value for the absorbance of a solution which only transmits 12% of the incident light.

(ii) Calculate the percentage of light transmitted for a solution with an absorbance value of 0.55.

(iii) Determine the value for the absorbance of a solution of an organic dye ($0.0007 \, \text{mol dm}^{-3}$) in a cell with a 2 cm pathlength if its absorptivity is $650 \, \text{dm}^3 \, \text{mol}^{-1} \, \text{cm}^{-1}$.

1.2.1 Measurement of Absorption

Beer's law as given by equation (1.9) is not directly applicable to chemical analysis. Neither I_0 nor I, as defined, can be directly measured in the laboratory because the solution to be studied must be held in some form of container and interaction will occur between the radiation and the walls of the container, thus producing a loss of intensity as a consequence of reflections or absorption. In order to correct for these effects the intensity of the beam transmitted through the solution is compared with the intensity of a beam passing through an identical cell containing only the solvent for the solution. We can then calculate an experimental absorbance that closely approximates the true absorbance of the solution by using the following:

$$A \cong \log \left(I_{solvent} / I_{solution} \right) \qquad\qquad (1.11)$$

We will use the experimentally determined absorbance defined in equation (1.11) throughout the rest of the Unit.

1.2.2 Preparation of Standard Solutions

The preparation of a suitable set of calibration solutions is common to most quantitative analytical procedures. It is necessary for you to ensure that you can undertake the calculations involved in the preparation of such solutions and that you are aware of the criteria which determine the accuracy of such solutions.

For the analysis described in the following section we need a series of solutions of potassium manganate(VII) containing manganese concentrations in the range from 0 to $30 \, mg \, dm^{-3}$.

In order to obtain the series of solutions to be measured, a stock solution was made by dissolving $0.0900 \, g$ of potassium manganate(VII) in water and making the volume up to $250.0 \, cm^3$ in a standard volumetric flask. The series of calibration solutions for the spectrophotometric analysis of Mn were then prepared by using the dilutions shown in the table below:

Solution	A	B	C	D	E
Volume of stock (cm^3)	2.00	4.00	6.00	8.00	12.00
Final diluted volume (cm^3)	50.0	50.0	50.0	50.0	50.0

Π Calculate the concentrations of the solutions A to E, expressing
the results in:

(a) $mg\,dm^{-3}$ of $KMnO_4$;

(b) $mg\,dm^{-3}$ of Mn.

(The relative atomic masses are K = 39.098, Mn = 54.938, O =
15.999.) Comment on the factors which will determine the
accuracy of the calibration solutions, mentioning the preferred
titrimetric method of standardising manganate(VII) solutions.

First, we calculate the concentration, in $mg\,dm^{-3}$, of the stock solution:

$$0.0900\,g\,KMnO_4 \text{ in } 250.0\,cm^3 = 4 \times 0.0900\,g\,dm^{-3}$$

$$= 0.3600\,g\,dm^{-3}$$

$$= 360\,mg\,dm^{-3} \text{ of } KMnO_4$$

(a) Allowing for dilution:

$$2.00\,cm^3 \text{ diluted to } 50.0\,cm^3 = 2/50 \times 360$$

$$= 14.4\,mg\,dm^{-3}$$

Thus, we can tabulate the $KMnO_4$ content for the solutions as follows:

Solution	A	B	C	D	E
Concentration ($mg\,dm^{-3}$)	14.4	28.8	43.2	57.6	86.4

(b) To calculate the manganese content we use the following ratio:

$$Mn/KMnO_4 = 54.938/(39.098 + 54.938 + 4 \times 15.999) = 0.3476$$

Hence, $14.4\,mg\,dm^{-3}$ $KMnO_4$ contain $0.3476 \times 14.4\,mg\,dm^{-3} = 5.00\,mg\,dm^{-3}$ Mn, which gives the following manganese contents:

Solution	A	B	C	D	E
Concentration ($mg\,dm^{-3}$)	5.00	10.00	15.00	20.00	30.00

It must be remembered that the accuracy of the derived concentrations will depend on the accuracy of the initial weighing and then, probably more importantly, on the accuracy of the volumes of solution measured and diluted.

Materials of the highest purity should be used for the preparation of any standard reagent solution (e.g. AnalaR $KMnO_4$ is of 99.9% purity) and the material should be carefully dried before use. Sodium oxalate (sodium ethandioate) is recommended as a primary standard suitable for checking the strength of the potassium manganate(VII) stock solution titrimetrically.

Note that the blank solution, i.e. one containing no Mn, in this case is a sample of the deionised water used to dissolve the potassium manganate(VII) and to make the dilutions.

1.2.3 Calibration Data for Manganate(VII) Analysis

Let us now look at some absorbance measurements on the solutions prepared in the previous section.

The first requirement for an analytical procedure which uses spectrophotometry is to have an absorbing system that is stoichiometrically related to the analyte to be determined. In this case, we have selected a manganese compound, $KMnO_4$, which has a stable and characteristic absorption spectrum in the visible region. Secondly, we must choose a suitable wavelength for measurement. This would normally be specified in the details of the analytical procedure being followed (see Chapter 3), or it would have been determined previously and noted. In this example, we have made no assumptions and have recorded the spectrum over the visible region. If we did not have a recording instrument we would make absorbance measurements near the expected position of λ_{max} and find which wavelength gave a maximum absorbance for one of the solutions.

We have recorded the transmittance spectra in the visible region of all of the solutions A to E prepared in the previous section. This allows you to see that the shape of the transmission spectrum does not change very much with concentration. All that occurs is that the transmittance values decrease as the solution colour becomes more intense, i.e. as the concentration of the $KMnO_4$ increases. It also gives us the opportunity to carry out some calculations for converting transmittance into absorbance.

For the solutions A to E the transmittance spectra shown in Figure 1.2c were measured over the visible region (700–400 nm) with the manganate(VII) solutions in 1 cm pathlength glass cells, using deionised water in the reference (or blank) cell.

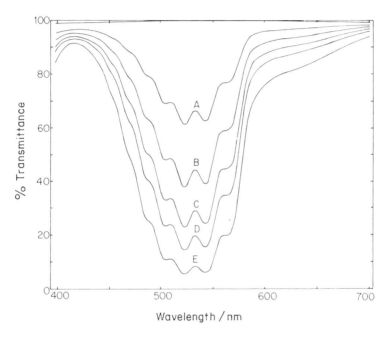

Figure 1.2c Transmittance spectra of potassium permanganate solutions; spectra from samples A to E represent increasing absorption (decreasing transmittance) in the green region of the spectrum

You can now use these spectra to perform a simple exercise as follows:

∏ (a) Read off, as accurately as you can, the wavelength of maximum absorption (λ_{max}) of the $KMnO_4$ solutions.

(b) Read off the transmittance values as % transmittance for each of the solutions at λ_{max}.

(c) From these transmittance values calculate the corresponding absorbance values.

(d) Plot your results as a calibration graph of absorbance against concentration of manganese.

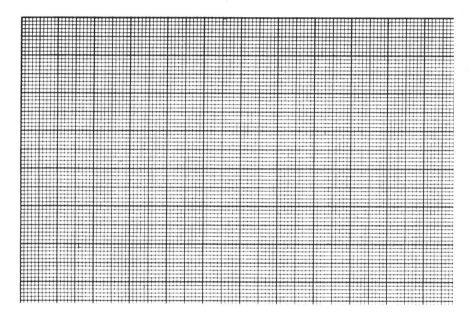

Did you get good agreement with the literature value of λ_{max} (i.e. 522 nm)? Do your points lie close to a straight line, and does this line pass through the origin? These last two criteria enable us to say whether the analyte obeys Beer's law over the concentration range which is being tested. In this case, Beer's law is obeyed.

You should have the following set of results:

Solution	Blank	A	B	C	D	E
Mn concentration $(mg\,dm^{-3})$	0.0	5.0	10.0	15.0	20.0	30.0
% Transmittance at 522 nm	100	61	38	23	15	6
I_0/I	1.0	1.6	2.6	4.3	6.7	16
$A = \log(I_0/I)$	0.00	0.20	0.41	0.63	0.82	1.22

You could, of course, have used linear regression to calculate the equation of the best straight line through your data points. If you are not familiar with this mathematical technique, you will come across it if you follow the Measurement, Statistics and Computation Unit in the ACOL Series.

Before we leave this section let us examine the significance of the molar absorptivity.

By rearranging Beer's law, equation (1.9), we obtain the following:

$$\epsilon = A/cl \qquad (1.12)$$

Since absorbance is dimensionless and hence has no units, by using values of the concentration and pathlength in units of $mol\,dm^{-3}$ and cm, respectively, we can calculate ϵ, which will have units of $dm^3\,mol^{-1}\,cm^{-1}$. This unit has been used for many years, but it is now becoming more common to find ϵ quoted in $dm^2\,mol^{-1}$, or the Système International (SI) unit of $m^2\,mol^{-1}$. Let us calculate some molar absorptivities and then examine their significance.

The molar absorptivity for $KMnO_4$ can be calculated from the data given above. We will use the data for solution C, as follows:

$$KMnO_4 \text{ concentration} = 43.2\,mg\,dm^{-3}$$

$$\text{absorbance} = 0.638$$

$$\text{cell pathlength} = 1\,cm$$

$$\lambda_{max} = 523\,nm$$

The relative molecular mass (M_r) of $KMnO_4$ is 158.032, and the molar concentration of solution C = $(43.2 \times 10^{-3})/158.032$

$$= 2.734 \times 10^{-4}\,mol\,dm^{-3}$$

$$\epsilon = A/cl$$

$$= 0.638/(2.734 \times 10^{-4} \times 1)$$

$$\epsilon = 2334\,dm^3\,mol^{-1}\,cm^{-1}$$

You could also have calculated the value for ϵ from your calibration graph simply by measuring the gradient of the straight line.

You will have probably realised that many analyses (and associated calculations of ϵ) are made at λ_{max}. This is because it gives maximum sensitivity in the analysis. It is, however, sometimes more convenient to work at other wavelengths, and so it is very important to quote the actual wavelength chosen. When absorption measurements are made at λ_{max} this is often indicated by adding the suffix to the symbol, i.e. ϵ_{max}.

In order to calculate the result in units of $dm^2\,mol^{-1}$, we need to express the pathlength in decimetres.

Applying this to our results above we have the following:

$$c = 2.734 \times 10^{-4}\,mol\,dm^{-3}$$

and $l = 0.1\,dm$ (i.e. 1/10 of the previous value)

giving $\epsilon = 0.638/(2.734 \times 10^{-4} \times 0.1) = 23\,300\,dm^2\,mol^{-1}$ (a value 10 times that which was obtained previously).

Note that in quoting these results the figures have been rounded off to give the final answer to three significant figures. This is reasonable in the context of routine spectroscopic analysis, which involves manual operations such as dissolution and dilution, and reading results from a graph. Variations of up to 2 or 3% are normal. Sometimes the relative molar mass of the analyte species is unknown and it is impossible to calculate the molar absorptivity. In such cases it is usual to calculate the following parameter:

$$E_{1\%}^{1\,cm}$$

which is the absorptivity representing the absorbance of a 1% solution in a 1 cm pathlength cell.

What is the analytical significance of the absorptivity? It will become apparent when you have done the following calculations.

SAQ 1.2b

Calculate the concentration, in units of mg dm^{-3}, of a solution of each of the two compounds A and B.

Compound	M_r	ϵ (dm^3 mol^{-1} cm^{-1})	Absorbance
A	250	1 000	0.10
B	250	100 000	0.10

What is the significance of the molar absorptivity in analysis?

1.2.4 Visual Comparison

In all *visual comparison methods* we aim to match the intensity of the colour of two samples, so let us start by putting the transmittance of the test solution, 1, equal to that of the standard solution, 2.

Hence $$T_1 = T_2 \tag{1.13}$$

Because $\log(1/T_1) = \log(1/T_2)$, the absorbances of the two samples are also equal and thus:

$$A_1 = A_2 \tag{1.14}$$

If we are comparing identical compounds, which obviously have identical spectral characteristics, then according to Beer's law the following relationship applies:

$$A_1 = \epsilon_1 c_1 l_1 = \epsilon_2 c_2 l_2 = A_2 \qquad (1.15)$$

and because $\qquad\qquad\qquad \epsilon_1 = \epsilon_2$

$$c_1 l_1 = c_2 l_2 \qquad (1.16)$$

This is the desired relationship for visual comparison techniques.

In the technique in which the dimensions of the two sample containers are identical ($l_1 = l_2$) and the samples are 'colour matched' then, as you would expect, $c_1 = c_2$.

In the alternative technique where colour matching is achieved by varying either l_1 and/or l_2, a simple calculation and instrument calibration is required, as we shall see later.

Although visual colorimetry provides one of the simplest and cheapest methods of quantitative analysis it suffers from the following limitations:

(a) it is restricted to the analysis of coloured materials;

(b) it is sensitive to even low concentrations of coloured impurities;

(c) it suffers from a lack of precision.

For visual comparison techniques to be successful the colour of the standard and test solutions (as determined by the shapes of the visible absorption spectra) must be identical. Only the intensity of the colour, i.e. the transmittance values, should vary.

1.2.5 Types of Visual Comparators

The simplest visual comparison technique is to place a series of standard solutions of varying colour intensity in a set of test tubes of identical dimensions. The unknown test solution is then placed in an identical test tube and compared in turn with each of the standard solutions by using the side-by-side viewing technique. The principal practical difficulty in using this technique is to ensure that the series of standard solutions covers the range of the test solutions with a sufficiently small increment in concentration to achieve the best realisable precision and accuracy. It is also necessary to prepare fresh standard solutions on a regular basis when determinations are to be carried out.

Use of a disc comparator such as the Lovibond Comparator (Figure 1.2d) eliminates the necessity of preparing standard solutions. With this equipment the set of standard solutions is replaced by a disc which holds a series of permanent coloured glass filters designed to simulate the colour of the standard solutions. The discs are held in a special viewing box into which a tube containing the coloured test solution is placed alongside a tube containing only the colourless blank or solvent, in line with the coloured filter.

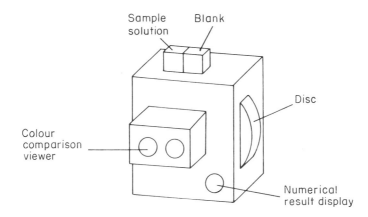

Figure 1.2d   Diagram of a disc comparator

One of the main disadvantages of this technique is that the simulated colour provided by the disc is not always a complete spectral match of the test solution. This results in the colour match being sensitive to the background illumination used for viewing, and to a lesser extent on the colour vision characteristics of the analyst. With the Lovibond discs the recommended illuminant is average daylight.

The application of direct comparison is best understood from the use of the traditional Dubosq-type colorimeter (Figure 1.2e), i.e. a visual comparator which avoids the requirement of a large series of standard solutions. This is achieved by allowing the depth of solution to be varied until a colour match is obtained with a single standard concentration. Therefore, Dubosq-type comparators have viewing tubes which can be inserted into both the unknown and a suitable standard solution and the relative lengths of the solutions viewed are then varied until a colour match is obtained.

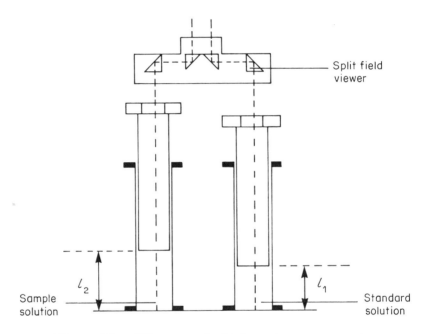

Figure 1.2e   Diagram of a Dubosq-type comparator

The relative depths are indicated on calibrated wheels which control the depth of the viewing tubes. To improve the precision of the colour-match point the light passing through the test and standard solutions are present in the eyepiece as two adjacent halves of the field of view. Thus if the standard of concentration c_1 requires a depth of l_1 to give the colour match to the test solution at depth l_2 then the required concentration of the test solution c_2 can be calculated from the following:

$$c_2 = \frac{c_1 l_1}{l_2} \qquad (1.17)$$

The following calculation illustrates this procedure using the results obtained from a Dubosq instrument.

Π A Dubosq-type comparator is being used to make measurements on permanganate solutions for the analysis of a steel for its manganese content. (The manganese is oxidised in solution by bismuthate or periodate.) A standard solution of 0.200 mg dm^{-3} Mn was set at a depth of 35.5 mm and the unknown was matched at a depth of 30.0 mm.

(a) Calculate the concentration of Mn in the unknown.

(b) If the λ_{max} of $KMnO_4$ is $2300 \text{ dm}^3 \text{ mol}^{-1} \text{ cm}^{-1}$, use Beer's law to calculate the absorbance and the % transmittance of the unknown solution at the pathlength quoted (relative atomic mass (A_r) of Mn = 54.938).

(c) Assuming a precision of $\pm 3\%$ in matching the transmittances calculate the possible error in the unknown concentration.

(a) Applying the Dubosq relationship, $c_2 = c_1 l_1 / l_2$:

$$\text{Mn concentration of unknown} = 0.200(35.5/30.0)$$
$$= 0.237 \text{ mg dm}^{-3}$$

(b) Applying Beer's law, $A = \epsilon c l$, we can calculate the concentration in units of mol dm^{-3}:

$$c = 0.237 \times 10^{-3}/54.938$$
$$= 4.23 \times 10^{-5} \text{ mol dm}^{-3}$$

Hence absorbance, $A = 2300 \times 4.23 \times 10^{-5} \times 3.00$

i.e. $A = 0.291$

since $A = \log(100/\% T) = 2 - \log \% T$

then $\log \% T = 2 - 0.291 = 1.709$

and $\% T = 51.2$

(c) Applying a $\pm 3\%$ relative error to the transmittance gives

$$51.2 \pm (3 \times 51/100) = 51.2 \pm 1.5$$

The range of values is 49.7 to 52.7 % T

This corresponds to a range of absorbance values of 0.304 to 0.278, or 0.291 ± 0.013.

Since absorbance $\propto$ concentration, the concentration is

$$0.237 \pm 0.007$$

$$\text{or} \quad 0.24 \pm 0.01 \, \text{mg dm}^{-3}$$

Merits of Visual Colorimetry

From the above discussion you should be able to list some of the advantages and disadvantages associated with visual colorimetry. How do you think the colorimetric determination of manganese in steel compares with say a volumetric method? Is the method quicker? Does it require a large amount of sample? Is the standardisation more difficult? These and many other similar questions must be asked before any judgement can be made on the relative merits of a particular type of analytical procedure. In the case of colorimetric methods the following advantages and disadvantages have been listed:

Advantages

(a) Colorimetric methods are usually rapid in comparison with volumetric and gravimetric methods.

(b) Often they require a minimum of sample preparation, with sometimes only dissolution and colour development.

(c) Usually only a small amount of sample is required — in many methods a few milligrams will suffice.

(d) The simplest of equipment is required, particularly in the case of the visual comparison technique.

(e) Highly trained technicians are not required, as even non-technical personnel can be trained to do simple visual comparisons of colour intensity.

Disadvantages

(a) The preparation of fundamental standards for colorimetry can be a problem, and for visual comparison methods these may have to be replaced at frequent intervals.

(b) Where coloured filters are used to simulate the standard a suitable source of illumination must be specified and be available.

(c) The presence of interfering ions can cause colour distortions and invalidate the visual comparison.

(d) The sensitivity of visual methods is not high; an absolute accuracy of $\pm 5\%$ may be expected routinely, which is much poorer than a good volumetric or gravimetric method of analysis.

1.3   BASICS OF ULTRAVIOLET/VISIBLE SPECTROMETERS

In the previous section we discussed visual comparison methods and the comparatively simple equipment required to carry out such techniques. If the chemical species that we are trying to analyse, however, is colourless but absorbs electromagnetic radiation in the accessible ultraviolet part of the spectrum we will need to use more complex instrumentation.

From what was discussed at the beginning of Section 1.2 about the criteria for applying absorption measurements for the determination of an analyte in solution, can you list the requirements for a spectrometer in order for it to be used for such measurements? The main requirement is a knowledge of the range of wavelengths over which the analyte solution absorbs. If the solution is coloured then we immediately know that it absorbs over the visible range and hence an instrument operating over the visible region is probably sufficient. If, however, you were expected to carry out an analysis of, e.g. quinine in a sample of tonic water, or a biochemical analysis, an instrument capable of measuring in both the ultraviolet and the visible regions of the spectrum is likely to be required.

Any instrument must, therefore, allow an appropriate wavelength to be selected which is suitable for the particular analyte being considered. The sample and reference or blank solutions must be placed in the light beam in such a way that the ratio of the transmitted radiation beams can be measured and, finally, the transmittance or, preferably, the absorbance value for the solution should be displayed and recorded. These requirements allow us to list the basic components of a UV/visible spectrometer, and secondly to look at the way these components are assembled in a typical instrument.

1.3.1 Instrument Components

There are five essential components required for most absorption spectrometers. These are as follows:

(a) A source or sources of radiation covering the required wavelength range.

(b) A means for selecting a narrow band of wavelengths.

(c) Facilities for holding the cells containing the sample solution and the blank in the radiation beam.

(d) A device or devices capable of measuring the intensity of the radiation beam transmitted through the cells.

(e) A display or output device to record the measured quantity in a suitable form.

The general arrangement of these components for a simple spectrometer is shown in Figure 1.3a.

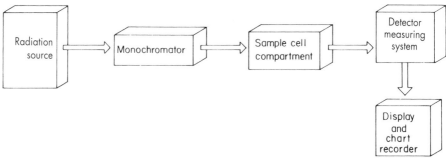

Figure 1.3a   Basic construction of a spectrometer

The Source

Two sources are generally used in UV/visible spectrometers which between them cover the whole range from 200 to 800 nm:

(a) For measurements above about 320 nm compact tungsten halogen sources in a quartz envelope are nowadays preferred to the older tungsten filament lamps because they give higher emission in the ultraviolet region. The older types of lamp were restricted to measurements above 360 nm. Tungsten halogen lamps contain a small quantity of iodine vapour within the quartz envelope housing the tungsten filament. The operating lifetime of such a lamp is about twice that of the older tungsten filament version because the iodine reacts with the tungsten vapour, which then sublimes off the filament to give WI_2. When molecules of this compound strike the hot tungsten filament decomposition occurs and tungsten metal is redeposited.

(b) For measurements below 320 nm a deuterium arc source is used as this emits a continuous spectrum below 400 nm. Special filters are often included in the optical path when a tungsten halogen lamp is being used below 400 nm. These are needed to reduce the amount of stray radiation reaching the detector, which may cause errors in the absorbance readings. The effects of stray radiation are considered further in Section 1.5.2 (see later).

Wavelength Selectors

You may remember that Beer's law assumes that monochromatic radiation will be used for absorption measurements. In practice, it is not possible to obtain monochromatic radiation by using the sources described previously. We therefore use radiation consisting of a limited, narrow continuous group of wavelengths called a *band*, which has a Gaussian-shaped distribution of wavelengths. A narrow bandwidth is required in order to enhance the sensitivity of the absorbance measurements.

Two types of wavelength selectors are usually employed:

(a) Filters — which provide only a limited wavelength selection.

(b) Monochromators — which allow continuous variation of the wavelength.

The effective bandwidth of a wavelength selector is an inverse measure of the quality of the device, i.e. a narrow bandwidth will represent better performance.

We will now consider the two types of wavelength selectors in more detail.

Figure 1.3b shows the performance of the two main types of filters, i.e. interference and absorption.

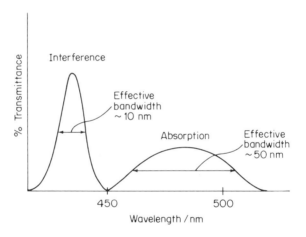

Figure 1.3b   Comparison of bandwidths for absorption and interference filters

Absorption filters work by absorbing part of the visible spectrum. They consist of coloured glass or a dye suspended in gelatine and sandwiched between glass plates. They have effective bandwidths that range from about 30 to 250 nm. Filters that provide the narrowest bandwidths also absorb a significant fraction of the desired radiation. They are cheaper than interference filters but their use is restricted to the visible region.

Interference filters consist of a transparent layer of a dielectric material such as calcium fluoride or magnesium fluoride which occupies the space between two plates of glass or another transparent material, as illustrated in Figure 1.3c. The thickness of the dielectric layer is carefully controlled and determines the wavelength of the transmitted radiation. Interference filters are available for both the visible and UV regions.

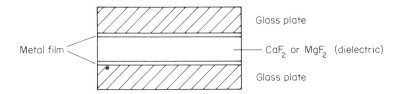

Metal film

Glass plate

CaF$_2$ or MgF$_2$ (dielectric)

Glass plate

Figure 1.3c   Construction of an interference filter

For methods where it is necessary to be able to vary continuously the wavelength of the radiation, monochromators are used. Monochromators for UV and visible radiation are all similar in mechanical construction in that they employ slits, lenses, mirrors, windows, and prisms or gratings. All monochromators contain an entrance slit, a collimating mirror or lens to produce a parallel beam of radiation, a prism or grating as a dispersing element and a focusing mirror or lens.

In most modern instruments, monochromation is likely to be based on the use of a diffraction grating which is arranged as illustrated in Figure 1.3d.

One disadvantage of diffraction gratings is the possibility of different spectral orders emerging from the exit slit with a given angular, and hence wavelength setting. Therefore, when a grating monochromator is set at 600 nm in its first-order spectrum, the same angle will allow some 300 nm radiation through from the second-order diffraction. A red filter is usually used in the lightpath when a grating monochromator is used above 600 nm in order to eliminate second- and higher-order transmissions. This problem did not occur with earlier instruments which used a prism for wavelength selection.

The quality of a monochromator can be measured by the resolving power, R, which describes the limit of its ability to separate adjacent

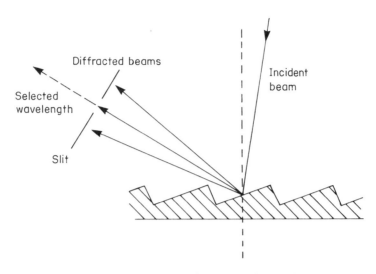

Figure 1.3d   Monochromation by a diffraction grating

images having a slight difference in wavelength. The resolving power is defined as follows:

$$R = \lambda/\Delta\lambda \qquad (1.18)$$

where λ is the average wavelength of the two images and $\Delta\lambda$ is their difference. The resolving power of typical bench-top monochromators ranges from 10^3 to 10^4.

The Sample Cell or Cuvette

Cells or cuvettes must be made of a material which is transparent to the radiation concerned, e.g. cells are made of silica for the ultraviolet/visible region and glass or plastic for the visible region. The optical windows of the cells are highly polished, flat and parallel and the lightpath between the inner surfaces is closely defined. The most commonly used cell is 10 mm in pathlength with a capacity to hold 3 to 4 cm³ of solution. A wide variety of pathlengths and cell volumes are, however, available. For small-volume cells, correct horizontal and vertical locations in the sample compartment are critical for ensuring that the specified cell pathlength is used.

The Detector

The function of the detector is to respond to radiation falling on the sensing surfaces and to provide an electrical signal which is proportional to the intensity of that radiation. Two main types of detectors are currently used in UV/visible spectrometers. Silicon photodiodes are now replacing the phototubes and photovoltaic cells incorporated in older instruments. Early silicon photodiodes had poor sensitivity below 400 nm but modern developments have improved their sensitivity so that they can now be used to below 250 nm.

For maximum sensitivity at low energies the photomultiplier tube is used in more expensive instruments. Photomultipliers have the advantage that they can be made to respond over the whole range from 190 to 950 nm. They need a high-voltage supply connected to the various dynodes within the tube that are used to amplify the initial electron emission from the photocathode surface. Many modern instruments now use diode array detectors.

Output Devices

In single-beam manual spectrometers produced up to the early 1970s the output device was invariably a meter of some form which either indicated the transmittance directly or was used as a null-point indicator in a potentiometric balancing circuit. The potentiometer control was usually calibrated both in transmittance and (non-linearly) in absorbance. Modern single-beam instruments are more likely to have a digital output linked directly to a microprocessor so that the display gives absorbance values directly, or can be calibrated in appropriate concentration units after standards have been measured. The more expensive recording instruments give a graphical chart record of absorbance against wavelength.

1.3.2   Instrument Layout

The above section gave some brief details of the components which go to make up a UV/visible spectrometer. You are not expected to remember all the details listed, but an understanding of the influence of the component characteristics on the reliability of the results that are produced should be the aim of all analysts. It is always a good idea to be able to at least read and understand the technical specification of an instrument.

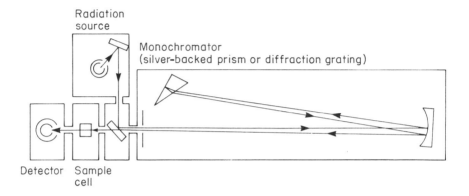

Figure 1.3e   Layout of a single-beam UV/visible spectrometer

For the present, let us now look at the optical layout of a single-beam spectrometer, as shown in Figure 1.3e. The position of the main components will readily be observed, with the additional front-surfaced mirrors being used to direct the light beams from one component to another. Can you indicate which of the five components is not shown in this figure? Yes, it is the output device.

When using a single-beam instrument it is necessary to take readings of the blank and the sample consecutively, which therefore relies on the characteristics of the instrument, in particular the source, remaining constant. Single-beam instruments are well suited for quantitative analysis where measurements are made at a single wavelength.

The arrangements in a recording instrument are similar, except that the measuring beam is split into two components and passed through a sample cell and a reference cell simultaneously. This is usually done with a rotating sector mirror, so that the detector receives an alternating set of pulses, where the intensity ratio of which gives the necessary transmittance ratio measurements. The signals are usually separated and converted to read-out values electronically. Thus, a double-beam instrument compensates for any source flicker or inhomogeneity. Its disadvantages are that it is more complex mechanically, is more expensive and has a lower sensitivity as measured by the signal-to-noise ratio.

To test your general understanding of the basics of a spectrometer a self-assessment question follows which merely asks for the identification of the components shown in the above figure, along with a question about the use of filters in the radiation path.

SAQ 1.3

Figure 1.3f below shows the optical components and layout of a typical recording spectrometer designed to operate over the wavelength range from 190 to 900 nm. Identify on the diagram the four principal components:

(i)   source;

(ii)   monochromator;

(iii)   sampling area;

(iv)   detector.

When the instrument is operated over certain wavelength ranges, filters are inserted into the optical path. Specify which of the filters, red or blue, is used:

(i)   at 780 nm;

(ii)   at 390 nm.

In each case, briefly explain the function of the filter used. What special precautions would you adopt to ensure the optimum instrument performance when measurements are being taken at 195 nm?

→

SAQ 1.3
(Contd)

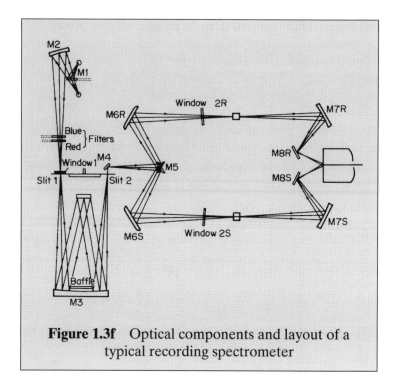

Figure 1.3f Optical components and layout of a
typical recording spectrometer

SAQ 1.3

1.4 SCOPE OF ANALYSIS BY ULTRAVIOLET/VISIBLE SPECTROSCOPY

You will by this time be aware that measurement of the absorption of UV/visible radiation provides one of the most widely used means of analysis of chemical and biochemical systems. Included under this heading are techniques ranging from the simplest colour test to the most recently developed multicomponent analyses using computer-controlled diode array UV/visible spectrometers.

The specific applications which were referred to earlier and which are currently in use in industry, hospitals and research departments and institutions can be classified in a variety of ways. Typical classifications would be as follows:

Colour tests
Visual colorimetry
Photometric analysis
Colour measurement and/or control
Kinetic spectroscopy (fast-reaction studies)
High performance liquid chromatography (HPLC) detection

The chemical and biochemical fields in which these methods are applied can also be classified in a number of ways, e.g.

Biochemical analysis
Enzymatic assays
Immunoassays
Pharmaceutical analysis
Trace-metal detection
Vitamin analysis
Quantitative organic analysis

You may be aware of other methods of classification for a group of techniques based on UV/visible absorption. It will already be evident to you that these fields of application are so diverse that it is only possible in a Unit of this present size to emphasise the common principles involved in using UV/visible spectroscopic methods. However, it is important for you to appreciate the diversity of the general technique and in this current section an attempt will be made to illustrate that diversity in a little more detail.

1.4.1   Colour Tests Today

The visual observation of a change in colour is so simple and straightforward that a number of regularly used analyses employ this principle. These include indicator papers for determining pH (acidity and alkalinity), starch-iodide papers for detecting oxidising agents, heat-test papers in the explosives industry, reagent papers and test sticks for medical diagnosis, and the latest 'dry-chemical' packs used for the clinical analysis of blood.

Such test papers or test kits can be designed to give both qualitative and quantitative information. Therefore, with the development of enzyme methods of analysis, and antigen–antibody complexes in the form of tests involving the observation of colour changes, clinical biochemists are converting what were previously specialist analyses into forms which are now suitable for use by doctors and nurses. The tests are designed to be robust, reproducible, highly specific and easy to carry out.

1.4.2   Visual Colorimetry (Using a Comparator)

The colour tests described above are a sub-set of the visual colorimetric methods of analysis, but in this section we will now look at some of the materials for which coloured glasses have been developed for use in the visual comparator of the Lovibond-type mentioned in Section 1.2.5. The list below also includes materials for which the colour is used as an indication of product quality, rather than being a measurement of chemical composition. Again the diversity of the substances should be noted.

Materials for which glass comparator colour standards are available include the following:

Air pollutants	Condensed milk	Honey	Rum
Alcohol	Cordials	Ink	Sand
Alloys	Cosmetics	Jellies	Sewage
Aniline	Cresols	Lacquers	Silk
Asphalt	Custard	Lard	Soap
Beer	Dental materials	Leather	Spirits
Benzene	Detergents	Malt	Sugars
Biscuits	Disinfectants	Mercury	Tobacco
Bitumen	Drugs	Metals	Varnish
Blood	Enamels	Milk	Vinegar
Butter	Fats	Oats	Water
Caramel	Fish	Paint	Wax
Carotene	Flour	Paper	Whisky
Celluloid	Foods	Pharmaceuticals	Wines
Cheese	Glass	Phenol	Wool
Chemicals	Glucose	Plastics	Yeast
Cider	Glue	Resin	
Coffee	Hair dyes	Rubber	

1.4.3   Biochemical Spectroscopy

Ultraviolet/visible spectroscopy has played an important role in the study of natural products from plant and animal sources. These investigations have involved the steps of isolation, characterisation, synthesis and biosynthesis, and then to biological modes of action; UV/visible spectroscopy has been of value in each of these stages for a wide range of biochemical materials. In the two-volume publication, *Biochemical Spectroscopy*, by R. A. Morton, some of the chapter titles give an indication of the range of natural products for which UV/visible spectroscopic studies have been undertaken. These include the following:

Carotenoids and related substances
Aminoacids, proteins and enzymes
Heterocyclic compounds, including nucleotides and nucleic acids

,, bile pigments and cytochromes
d related substances
, ιαιιιιιns and coenzymes
Certain antibiotics and medicinal substances

For example, much of the work on steroids carried out in the USA by Professor Robert Woodward involved using UV/visible spectrometry as a basis for assigning chemical structures to compounds often differing only by the location of a carbon–carbon double bond or of a carbonyl group. He was able to predict the wavelengths at which maximum absorptions would occur, simply from a knowledge of the component chemical groups in the molecules.

Although many of these natural products have characteristic UV/visible spectra when pure, it is not always easy to use spectroscopic methods of analysis when they are present in complex mixtures. However, the analysis of specific components in complex mixtures, such as those of the food industry, can be achieved by enzymatic assays. Such assays are often based on the measurement of an increase or decrease in the absorbance of nicotinamide–adenine dinucleotide (NAD) at 340 nm (UV method), or alternatively its reduced form (NADH), can be oxidised in the presence of a tetrazolium salt to yield a coloured dye (colorimetric method). Enzyme assays of this type have been developed for such substances as amino acids, carbohydrates and organic acids.

A typical procedure for the determination of sucrose and glucose in foodstuffs is given in Chapter 3 of this Unit and includes the enzymatic hydrolysis of sucrose to glucose and fructose prior to determination of the glucose by measuring the quantitative amount of the reduced nicotinamide–adenine dinucleotide phosphate (NADPH) produced in the process.

Direct photometric analysis of biochemical materials is now less frequently carried out as it has been replaced by these enzymatic methods, and more recently by immunoassay methods. However, complexometric methods involving colour-forming reagents are still used for determining certain common metal ions such as calcium, magnesium and iron, as will be discussed in Chapter 2 of this Unit.

1.4.4 Ultraviolet/Visible Monitors

Special small-volume, flow-through UV sample cells have been developed for high performance liquid chromatography (HPLC) systems. These enable the continuous monitoring of eluted components from the chromatographic column. The detection of individual species depends on their abilities to absorb the radiation employed in the detector. When analyte species do not absorb radiation at the operating wavelengths of the detector it may be possible to detect them by incorporating an absorbing centre in the molecule by an appropriate chemical reaction. Such procedures are termed 'chemical derivitisation' or labelling reactions. A common example is the derivitisation of amino acids using dansyl chloride, as follows:

SAQ 1.4	Indicate which of the following statements are *true* and which are *false*:
	(i) Quantitative colorimetric analysis requires the use of a light-measuring instrument.
	(ii) The eye is as good as an instrument for detecting colour changes.
	(iii) All colorimetric methods of analysis have been developed for trace levels.
	(iv) Certain biochemical methods of analysis can be designed to be highly selective, even though they all involve the measurement of change in the UV spectrum of nicotinamide–adenine dinucleotide (NAD).

→

SAQ 1.4   (v) The use of UV/visible spectroscopy in the
(Contd)   monitoring of the separation of mixtures by
                      high performance liquid chromatography
                      (HPLC) is limited to those components which
                      show strong absorption above 220 nm.

1.5 LIMITATIONS OF BEER'S LAW

Earlier in this chapter we showed that there was a linear relationship
between absorbance and concentration, i.e. Beer's law. Few
exceptions are found to the generalisation that absorbance and
pathlength are linearly related but deviations from direct
proportionality between absorbance and concentration when the
pathlength is constant are frequently encountered. These deviations
result in calibration graphs which are not linear, with the departure

from non-linearity being particularly severe at high concentrations. Some of these deviations are fundamental and represent *real limitations* of the law. Others occur as a consequence of the manner in which the absorbance measurements are made, or as a result of chemical changes occurring which are associated with concentration changes; the latter two are sometimes known as *instrumental deviations* and *chemical deviations*, respectively. We will now consider each of these in turn.

1.5.1 Real Deviations

As a general rule, concentration effects are not usually encountered at concentrations $<0.01\,mol\,dm^{-3}$. Above this value, however, refractive index changes and the perturbing effects of intermolecular interactions, or of ionic species, on the charge distribution of the absorbing species can affect the value of the molar absorptivity and give rise to either positive or negative deviations. In cases where the absorbing compound is involved in a concentration-dependent chemical equilibrium such as dimerisation, marked deviations will be observed if the spectral characteristics of the monomer are significantly different.

Some exceptions to Beer's law, even at concentrations $<0.01\,mol\,dm^{-3}$, are observed for certain large organic ions or molecules. For example, the molar absorptivity at 436 nm for the methylene blue cation increases by about 90% as the dye concentration is increased from 10^{-5} to $10^{-2}\,mol\,dm^{-3}$. Even below $10^{-6}\,mol\,dm^{-3}$ strict adherence to Beer's law is not observed.

1.5.2 Instrumental Deviations

Instrument and Spectral Bandwidths

In a spectrometer the ideal monochromator would enable the selection of radiation of any single wavelength within a given range, but in practice this situation is never achieved and instead of a single wavelength the 'monochromated' beam actually has a spread of values on either side of the required value. This instrument bandwidth corresponds to the width of the transmitted band at half the maximum transmittance value.

Modern UV/visible instruments usually have adjustable bandwidth settings, typically in the range 0.5 to 8 nm. The higher bandwidths are used where samples are strongly scattering or absorbing, i.e. where the signal is low. Wide bandwidths do have an effect on the absorption curve profile, and this needs to be appreciated (see below). Conversely, if the fine structure in a spectrum is to be resolved then narrow bandwidths are required, subject to signal-to-noise requirements.

As indicated above, the magnitude of the instrument bandwidth which is selected for measurement can influence noise levels, spectral resolution and absorbance values. For quantitative analysis it is the influence on absorbance values which is of most importance. The effect is a function of the ratio of the instrument bandwidth (IBW) to the spectral bandwidth (SBW) of the sample which is being measured. Typical factors for the effect on absorbance values are as follows:

IBW/SBW ratio	0.1	0.25	0.5	1.0	1.5
Absorbance factor	1.0	0.96	0.87	0.66	0.55

This table shows that, for a given instrument bandwidth, the effect is most noticeable for materials with a sharp absorption peak in the UV/visible region.

The data also show that the instrument bandwidth needs to be a factor of 10 less than the spectral bandwidth of the species being measured if accurate absorbance values are to be obtained. This is important because if the IBW/SBW ratio is not less than 0.1, Beer's law will not be obeyed and negative deviations will therefore occur.

Experimentally, deviations from Beer's law resulting from the use of a polychromatic beam are not appreciable provided that the radiation used does not encompass a spectral region in which the absorber exhibits large changes in absorption as a function of wavelength. This is illustrated in Figure 1.5a where measurements are made at the peak maximum and on the sloping side of the absorption peak.

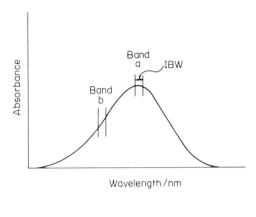

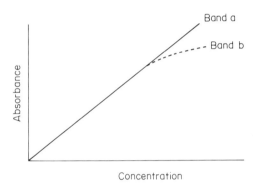

Figure 1.5a Effect of polychromatic radiation if the absorbance measurements are not taken at λ_{max}

Stray Radiation

When we set a monochromator to pass radiation of a particular wavelength we wish to prevent radiation of higher and lower wavelengths from reaching the detector. Much of the stray radiation may not have passed through the sample. If the instrument has a stray light level of 0.1% we can only stop 99.9% of the unwanted radiation from reaching the detector. Let us find out how good this and other levels are and the effect that stray light has on the absorbance values we obtain.

We have already considered an equation relating absorbance to

% transmittance, namely:

$$A = \log(100/\% T) \qquad (1.19)$$

Let us modify this to enable us to calculate the effects of stray light. If S is the % stray light the equation then becomes:

$$A = \log[(100 + S)/(\% T + S)] \qquad (1.20)$$

This tells us that the absorbance values are reduced by the presence of stray radiation and that this reduction becomes more significant at high absorbance values. This is illustrated in Figure 1.5b, which shows the apparent negative deviation from Beer's law as a result of stray radiation.

At high concentrations and longer pathlengths stray radiation can cause deviations from the linear relationship between the absorbance

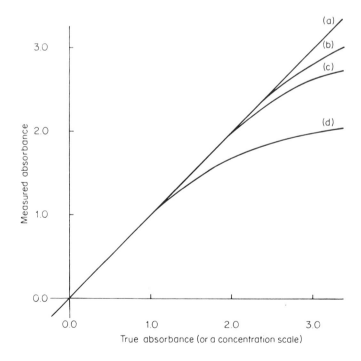

Figure 1.5b   Deviations from Beer's law due to stray radiation, S:
(a) 0; (b) 0.05; (c) 0.1; (d) 1.0%

and pathlength. In addition, the presence of stray radiation can also give rise to distortions in the shape and position of the absorption bands (Figure 1.5c).

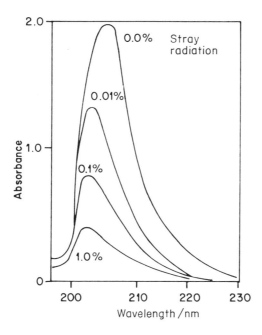

Figure 1.5c Stray radiation effects on the absorbance spectrum of maleic acid in ethanol

The main source of stray radiation in most spectrometers is usually the dispersing element in the monochromator, although scattering from other optical surfaces and deposited dust increases markedly with the age of the instrument. The effect of stray radiation is most noticeable in the low-wavelength region, particularly below 220 nm, where the source energy and detector sensitivity are lowest and also where the transmission characteristics through the optics are decreasing. The stray radiation characteristics are usually quoted as a percentage of the transmittance signal, at say 220 nm, and for a new instrument this should be less than 0.1%.

1.5.3 Chemical Deviations

Of particular interest to the analytical scientist is the effect of pH on chemical equilibria, which is made use of so often. For example, we choose a visual indicator for monitoring acid–base titrations whose colour is very dependent on the pH of the solution.

In order to illustrate the effect of pH on equilibria, let us look at the chromate/dichromate equilibrium. It is of importance in ultraviolet/visible spectroscopy because standard solutions of potassium dichromate are used for checking the accuracy of the absorbance scale of instruments.

The equilibrium can be represented as follows:

$$Cr_2O_7^{2-} + H_2O \rightleftharpoons 2CrO_4^- + 2H^+$$

The visible absorption spectrum of a standard solution of $K_2Cr_2O_7$ at different pH values is shown in Figure 1.5d.

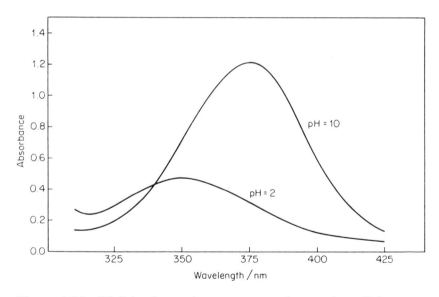

Figure 1.5d Visible absorption spectrum of potassium dichromate
$(3 \times 10^{-3} \, mol \, dm^{-3})$ recorded at pH 2 and 10

∏   Examine Figure 1.5d and then suggest two possible ways in which you might overcome the problems of the pH-dependent equilibrium when preparing a calibration curve for the determination of chromium(VI) using standard solutions of potassium dichromate.

One of the ways is chemical and the other spectral.

The chemical method is to simply buffer all solutions. If the pH is buffered at say less than pH 3 you would make all your measurements at 348 nm. An alternative, spectral method would be to make all measurements on unbuffered solutions at the wavelength where the curves overlap, known as the *isobestic point*, because the absorbance is independent of pH at this wavelength—340 nm in this case.

Isobestic points occur any time that we have two absorbing species which are interconvertible, as are chromate and dichromate in our example. The absorbance at the isobestic point is independent of the position of the equilibrium and only depends on the total amount of substance which is present.

A similar effect is observed with acid/base indicators. For example, the colour change associated with an indicator HIn arises from shifts in the following equilibrium:

$$HIn \rightleftharpoons H^+ + In^-$$
$$\text{Colour 1} \qquad\qquad \text{Colour 2}$$

If we plot absorbance versus concentration at the λ_{max} values for the two coloured species a non-linear graph, such as that shown in Figure 1.5e,

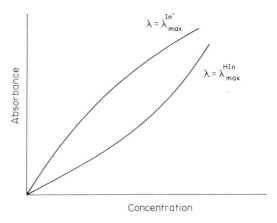

Figure 1.5e   Deviation from Beer's law for an acid/base indicator, HIn

would be obtained. We will return to the problems of using non-linear calibration graphs in the next section.

1.5.4   Non-linear Calibration Graphs

What are the problems we face when the Beer's law plot is non-linear? Can we use non-linear calibration graphs and obtain accurate and precise analytical results?

Let us begin to answer these questions by drawing our own curve. Figure 1.5f is an incomplete calibration curve in which six data points have been plotted. Complete the figure by drawing the best curved line to fit the points.

If you managed to draw a perfectly smooth narrow line by having

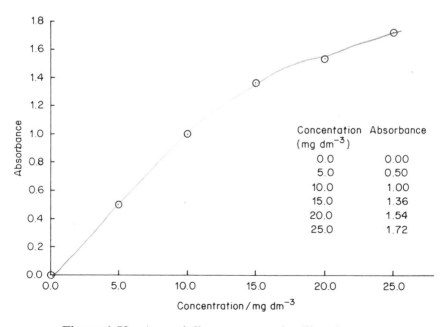

Concentration $(mg\ dm^{-3})$	Absorbance
0.0	0.00
5.0	0.50
10.0	1.00
15.0	1.36
20.0	1.54
25.0	1.72

Figure 1.5f   A partially constructed calibration curve

available a sharp pencil and a 'flexicurve' ruler, you would have been left with the difficulty of deciding exactly *where* to draw the line.

∏ Would more data points in the region of 10 to 25 mg dm^{-3} have been helpful?

Yes. It is not easy to accurately define the curve in this region. We could also have done better by calculating the best curve using a third-order polynomial and you would have done this if you had access to a microcomputer with appropriate software. Even so, a calculated curve would also leave something to be desired because of the limited number of data points that are available in this case.

∏ A solution of unknown concentration gave an absorbance of 1.45. Use your calibration curve to determine the concentration of this solution. Do you think your answer is likely to be inaccurate?

Probably, because it depends very strongly on the exact positioning of your curve and on how accurately you interpolated between the absorbance and concentration axes. My value is 17.1 mg dm^{-3}.

∏ It is obviously bad practice to use the region of a calibration graph where the curvature is pronounced. If you were faced with this problem for a solution giving an absorbance corresponding to a particularly non-linear region of the curve, what should you do?

Accurately dilute the solution so that the results fall in a linear region of the graph. Calibration curves for analyte species with small absorption coefficients can often deviate from Beer's law even at fairly low concentrations, and dilution of the sample is no longer advantageous. Are we justified in using the curve? When the non-linearity is not severe the graph itself can be drawn and then read with accuracy. Are we justified, therefore, in using this curve when the basic law of quantitative spectrometry is not strictly obeyed? That depends on whether or not the deviation from this law, which is evident from measurements on the standard solutions, is exactly the same for the solution of the sample that is being analysed.

We started this section by asking two questions. The first is rephrased in the following self-assessment question.

SAQ 1.5

> List two of the problems that might occur when calibration data do not obey Beer's law over the concentration range of interest.

The second question was — can we use non-linear calibration graphs and obtain accurate and precise analytical results?

You should be able to do so if you are careful. You are now aware of the problems and how they may be reduced to acceptable levels.

1.6   SPECTRA–STRUCTURE RELATIONSHIPS

So far, we have dealt with the nature and measurement of UV/visible spectra but have not considered in any depth how these spectra arise. What structural features of molecules give rise to absorption of radiation at various wavelengths and hence lead to different colours being observed for different compounds?

✓ This absorption occurs due to the fact that all molecules possess electrons which can be excited (raised to a higher energy level). Many of these electrons are excited by radiation of ultraviolet or visible wavelengths, but others are only excited by radiation within the vacuum-ultraviolet region.

1.6.1 Electronic Spectra

If only transitions in electron energy levels were involved, the UV/visible spectra for all compounds would consist of fairly sharp lines, i.e. very narrow absorption bands.

The energy difference between the electronic energy levels is given by the following equation:

$$\Delta E = h\nu \qquad (1.21)$$

In its simplest form this can be represented diagrammatically as shown in Figure 1.6a, in which the absorption of energy leads to electrons, initially in the ground state, moving to an excited state (an energy level of greater energy than the ground state).

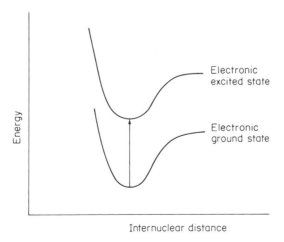

Figure 1.6a   Electronic transition energy diagram

However, as you have seen, most of the spectra are actually very broad smooth curves, and not sharp peaks. This is because any change in the electronic energy is accompanied by a corresponding change in the vibrational and rotational energy levels. You may already know that vibrational and rotational energy changes by themselves give rise to infrared absorption spectra. However, when they accompany ultraviolet/visible absorptions a large number of possibilities exist within each electronic state and the individual absorption bands normally become very broad.

The energy transition illustrated in Figure 1.6a does not represent the full story, as the change will actually be from a vibrational energy level in the electronic ground state to one of several vibrational levels within the excited state. You can see this more clearly in Figure 1.6b.

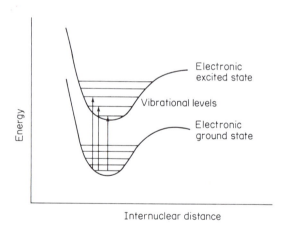

Figure 1.6b   Electronic transitions between vibrational energy levels

In some instances the fine structure produced by the vibrational levels may be observed in the ultraviolet/visible spectra of compounds such as benzene and toluene in the gaseous state. However, this is not normally observed with solutions and substances in the liquid state, although benzene does still produce a spectrum with a certain number of sharp well defined fine-structure peaks.

1.6.2   Structure and Energy

A variety of energy absorptions is possible, depending upon the nature of the bonds within a molecule. For example, electrons in organic molecules may be in strong σ bonds, in weaker π bonds or be non-bonding (n). When energy is absorbed all of these types of electrons can be elevated to excited antibonding states which can be represented diagrammatically as shown in Figure 1.6c, with the antibonding states being represented with an asterisk as σ^* and π^*.

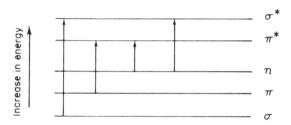

Figure 1.6c   Bonding and antibonding energy transitions

Most σ to σ^* absorptions for individual bonds take place below 200 nm in the vacuum-ultraviolet region and compounds containing just σ bonds are transparent in the near-ultraviolet/visible region. $\pi \rightarrow \pi^*$ and n $\rightarrow \pi^*$ absorptions occur in the near-ultraviolet/visible region, and result from the presence in the molecules of unsaturated groups known as *chromophores*. These are dealt with more fully in Chapter 5, but at this stage you should know that chromophores have characteristic molar absorptivities and absorb at fairly well defined wavelengths. Some typical chromophores are listed in Table 1.6.

The wavelengths of these characteristic absorptions and their molar absorptivities are often greatly changed due to the presence of other chemical groups in the molecular structure. It is found that groups such as –OH, –NH$_2$, and halogens, which all possess unshared electrons, cause the normal chromophoric absorptions to occur at a longer wavelength (i.e. displaced towards the red end of the spectrum), and with an increase in the value of the molar absorptivity. Groups which cause this change are known as *auxochromes*.

Table 1.6   Typical chromophores

Chromophore	Typical compound	Electronic transition	Characteristic values	
			λ_{max} (nm)	ϵ (m² mol⁻¹)
>C=C<	Ethene	$\pi \to \pi^*$	180	1300
>C=O	Propanone	$\pi \to \pi^*$	185	95
		$n \to \pi^*$	277	2
⬡	Benzene	$\pi \to \pi^*$	200	800
			255	22
—N=N—	Azomethane	$n \to \pi^*$	347	1
—N=O	Nitrosobutane	$n \to \pi^*$	665	2

This means that compounds possessing several chromophores and auxochromes are likely to be coloured. You will also find that the greater the degree of conjugation in the molecule (by this we mean the number of alternate double and single bonds) the longer the wavelength at which the ultraviolet/visible absorption will occur. Substances such as carotenes, which contain 11 alternate double and single bonds, appear bright red in colour because the extensive conjugation causes them to absorb strongly in the blue part of the visible region of the spectrum.

1.6.3   Absorptions in Inorganic and Organometallic Compounds

You will be well aware that many inorganic compounds such as potassium permanganate and sodium dichromate are very highly coloured, although they do not possess the unsaturated, conjugated structures which are characteristic of many organic compounds. The colour is due to transitions occurring in the energy levels of the d-electrons in the transition metals, i.e. in the manganese atom of the potassium permanganate and the chromium atom of the sodium dichromate.

Where transition metals are also linked to organic ligands in organometallic compounds very intense colours can arise as a result of d → d, n → π^* and $\pi \to \pi^*$ transitions all taking place. Another type of transition which affects the colour of these compounds is known as charge transfer, in which an electron occupying a σ- or π-orbital in the ligand is transferred to an unfilled orbital of the metal, and vice versa. The strong colours resulting from these various effects means that the formation of organometallic compounds is a useful procedure to adopt in trace-metal analysis, as even very small quantities of the metals can be made to produce compounds with high molar absorptivities.

You will find the topics of structure, conjugation and colour dealt with in more detail in Chapter 5.

Summary

For many years the measurement of light absorption and transmission has served as a basis for the measurement of concentrations of substances in solution. As a result of this a series of mathematical equations have been developed relating the degree of light absorption with the effective solution pathlength, the nature of the substance, and its concentration. Special instruments have been manufactured which enable the ultraviolet and visible radiation to be measured with a high degree of accuracy at different wavelengths.

Objectives

On completion of this chapter you should now:

- understand the importance of colour measurements in chemical analysis;

- be able to carry out calculations based upon Beer's law;

- have acquired a knowledge of the functions of the various components of a spectrometer;

- understand why deviations from Beer's law may occur;

- appreciate the importance and scope of UV/visible spectroscopy in a wide range of applications.

2. Quantitative Methodology

In the first chapter we dealt with the basic theory of spectroscopy as applied to quantitative analysis using ultraviolet and visible radiation. In this present chapter we will discuss in more detail the methodology for obtaining quantitative information about an analyte.

Probably you will have already used a spectrometer to carry out at least one quantitative analysis on a coloured solution. Let us suppose that you are familiar with the method for the determination of iron in water, based on the formation of the red-orange complex of iron(II) with 1,10-phenanthroline (the most often used complexing reagent for iron(II) when present in low concentrations). The method of analysis used would normally be given in the standard methods for such an analysis in, for example, the water industry.

∏ Can you suggest what details would be specified in a typical procedure involving a spectrometric analysis of an inorganic component in a sample of tap water?

You can check your answers against the following list:

(a) the amount of material to be used;

(b) the detailed method for forming the coloured complex — this could include the amounts of complexing agent, reducing agent and buffer solution, and the time to wait before the solution is ready for measurement;

(c) the preparation of a set of calibration solutions;

(d) the wavelength selected for the measurement and the pathlength of the cell to be used;

(e) the method of calculating the analytical results for the experimental measurements.

How many of these points did you list? It is not exhaustive and so you may have listed some others. If so, well done.

∏ What other steps would need to be specified for the determination of an inorganic component in an organic matrix such as a food product?

The organic matrix may well interfere with the determination and so it would need to be removed. Two additional steps to be taken are, therefore:

(a) removal of the organic matter, either by wet oxidation or by ashing;

(b) preparation of a solution of the residue by an acid dissolution procedure.

Some examples of the detailed specifications of spectrometric analysis are given in Chapter 3 of this Unit, but one object of this present chapter is to consider some of the factors which are important whenever a spectrometric analysis in the UV/visible region is attempted.

Returning to the determination of the iron content of water, some questions that might be asked include the following:

Is 1,10-phenanthroline the best reagent to use for iron?

Why is a reducing agent needed?

Why do we add the buffer solution?

If time is important, does that mean that the colour of the complex is unstable?

Do we always make the measurements at the wavelength of maximum absorption?

Can we use plastic cells to hold the solution to be measured?

These and other questions are usually considered by scientists who test proposed analytical methods before they are adopted as standard methods, as for example in the water or pharmaceutical industries.

Therefore, when a sample is to be analysed by UV/visible spectroscopy, the method of sample preparation, solution conditions of measurement, and instrumental parameters to be used will have been carefully considered and standardised during the design of the analytical procedure.

In the ideal situation the desired analyte species would be easily isolated, and converted to a highly absorbing form which possesses a characteristic absorption band within the range of the available instrumentation. The absorbing species would be stable and unaffected by solution conditions. Interference from other components in the solution should not occur. The system should obey Beer's law over a wide concentration range and the calibration graph should be reproducible and insensitive to small changes in instrumental characteristics.

In practice, however, ideal conditions rarely exist, and the purpose of this chapter is to examine the criteria which enable the optimum conditions to be selected for reliable spectrosvcpopic analysis.

2.1 SOLUTION PREPARATION, SOLVENTS AND CELLS

The appropriate treatment for obtaining a solution of the desired analyte is determined by the nature of the sample, the component to be determined, the other constituents present, the desired accuracy and the time available. For our purposes we will assume that we have a solution of the absorbing component, with the absorption either being a property of the analyte itself or of a chemical derivative of that analyte.

2.1.1 Stability and Solubility

In order to be suitable for spectroscopic analysis in the UV/visible region of the spectrum the analyte solution should possess the following properties:

(a) The absorbing species should have a stability for a sufficient enough time to permit accurate absorbance measurements to be made. Obviously, if the absorption is changing as you make the measurements then the result will have poor accuracy. Instability can arise as a result of many factors, such as air oxidation or photochemical decomposition, or be due to solution conditions such as solvent, pH and temperature. In some cases, little can be done to improve the stability, and it is, therefore, very important that the procedures used and the interval between the measurements are both carefully standardised so that any deterioration is constant and reproducible conditions are employed.

(b) Colloidal or insoluble material must not develop as a result of slow hydrolysis or some other type of reaction with the solvent.

Reactions yielding a colloidal system or suspension are difficult to control and are generally susceptible to the presence of electrolytes and other constituents. Stabilising additives such as agar and gelatine are only partially satisfactory. If the product is insoluble it can sometimes be extracted in another solvent.

The principal effect of the presence of colloidal or suspended material is an apparant increase in absorbance due to the scattering of the radiation by the colloidal or suspended particles. This effect is strongly wavelength dependent and so it is much more noticeable in the ultraviolet region, i.e. below about 300 nm, than it is in the long-wavelength region, i.e. above about 500 nm.

2.1.2 Choice of Solvent

From the above discussion it is apparent that the important characteristics of the solvent are as follows:

(a) good solubilising power;
(b) stable interactions with the absorbing species.

In addition, from the optical point of view, the solvent should not itself absorb in the region of measurement and should be of consistent purity.

Water is the cheapest and most transparent solvent and therefore it is commonly used for water-soluble substances. Unfortunately, only a small proportion of organic compounds are soluble in water and so organic solvents are needed. A practical difficulty with using water as the solvent in closed cells is its tendency to give rise to air bubbles which cause errors due to scattering of the radiation passing through the cell. The best remedy is to use freshly boiled water which contains little or no dissolved air. Water which has been distilled or deionised is generally of sufficient purity, but water which has been stored for long periods in plastic containers may contain small amounts of UV-absorbing impurities.

Alcohols, such as methanol, ethanol and propan-2-ol, also have good solubilising power and, if carefully purified, they will have good transparency at wavelengths below about 250 nm. Hexane and cyclohexane are also transparent in the UV region, provided that any traces of benzene or other aromatics have been removed before use by passing through a silica column or by shaking with sulfuric acid, or alternatively, Spectroscopic Grade solvents can be used.

Solvents such as trichloromethane, tetrachloromethane and the linear and cyclic ethers are commonly used in solvent-extraction procedures, particularly for metal complexes, but they have limited transparency in the UV region below about 300 nm. Diethyl ether is also unsuitable because of its high volatility.

The effective cut-off wavelengths in the ultraviolet region for a range of solvents are given in Table 2.1.

In addition to considering the solubilising power and transparency of a solvent it is also important to consider the safety aspects of using a particular solvent. Pyridine and dimethyl sulfoxide, for example, have very unpleasant smells and are highly toxic, and so should always be used in a fumehood, while benzene is a known carcinogen.

2.1.3   Sample Cells (Cuvettes)

Cells for the visible region may be made of glass (or transparent plastic if aqueous solutions are being used), but for the UV region

Table 2.1　Cut-off wavelengths for common solvents, with values given at the transmittance falls to 25% ($A=0.602$); measured by using water as a reference in 1 cm cells

Solvent	λ (nm)
Hexane	199
Heptane	200
Diethyl ether	205
Ethanol	207
Propan-2-ol	209
Methanol	210
Cyclohexane	212
Acetonitrile	213
Dioxan	216
Dichloromethane	233
Tetrahydrofuran	238
Trichloromethane	247
Tetrachloromethane	257
Dimethyl sulfoxide	270
N,N-dimethylformamide	271
Benzene	280
Pyridine	306
Propanone	331

below 330 nm quartz or fused silica cells must be used. A cell is specified by its type, material of construction, pathlength and dimensional tolerances.

Some typical cells are shown in Figure 2.1. These cells can be classified as:

Sampling cells

Flow cells

Rectangular cells

Sampling cells are fitted with tubes so that they can be filled and emptied by pressure or vacuum without having to be removed from the

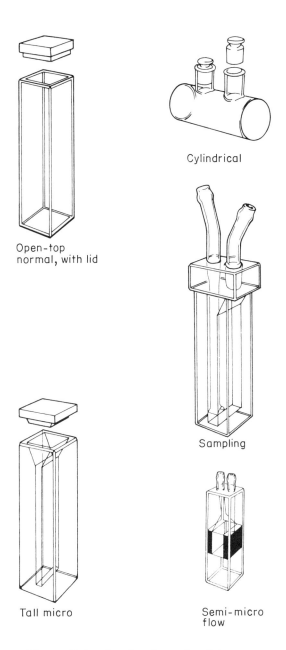

Open-top
normal, with lid

Cylindrical

Sampling

Tall micro

Semi-micro
flow

Figure 2.1 A selection of cell types

instrument. Usually they must be emptied as completely as possible before refilling. This type is often used when a spectrometer has been adapted for a large number of identical determinations involving stable solutions. Many instrument manufacturers offer accessories, often called *sippers*, for automatic emptying and filling of sampling cells.

Flow cells are intended for continuous flow operation and are designed so that each sample completely displaces the preceding one. They may be used with a continuously varying sample, as in a chromatography detector unit of the HPLC type or an autosampler.

Rectangular cells are simple containers which are filled and emptied manually and may or may not be removed from the instrument before refilling. The popular rectangular 1 cm cells are of this type.

These common types of cells can be obtained in three grades, depending on the optical quality and dimensional tolerances. Grade A cells are of the highest quality and pathlength tolerances are less than 0.1%, e.g. 10 ± 0.01 mm. Grade B cells are good quality cells for routine use with tolerances up to 0.5%, e.g. 10 ± 0.04 mm. Grade C cells are the cheapest and are used as disposable cells for large-scale routine work or in elementary teaching laboratories; dimensional tolerances are often as high as 3%, e.g. 10 ± 0.3 mm.

Even high quality cells differ a little from each other, and before embarking on accurate quantitative work it is normal to check sets of cells in order to have at least a matching pair — one to be used for the blank or reference solution and the other for the sample(s). The matching should be carried out to ensure that they have the same baseline transmission characteristics, as well as being of the same materials and possessing identical dimensions.

When you next carry out an analysis in which you use a pair of cells, spend some time making sure that you use them correctly. The following comments may help you in this respect.

When the instrument has been set up with the appropriate wavelength selected, you should set the absorbance reading to 0.00 (or the transmittance to 100%) *before* inserting the cells.

Wash the cells thoroughly, and fill both of them to within about 1 cm of the top with the blank solution. Ensure that the outer surfaces are clean and dry and that there are no finger marks or smears. Check that no air bubbles start to form on the inside surfaces of the cell. This will sometimes occur if a cool liquid starts to warm up in the cell and the solubility of the dissolved gases decreases.

Now insert you cells into the instrument and check the reading. Experience shows that cells are rarely matched and a small (but non-zero) reading is observed. If this is less than about 0.02 absorbance, then set the absorbance reading to zero. If the reading is >0.02 it is very likely that something is wrong and you should check for dirt and air bubbles. It is bad practice to 'back-off' (i.e. zero) such a big discrepancy.

SAQ 2.1

A biochemical enzymatic analysis is being carried out at 340 nm by spectroscopic measurements. Indicate which of the following would result in a large (L) and which would result in a small (S) effect on the measured absorbance.

(i) The sample becomes cloudy due to poor solubility.   L/S

(ii) The sample is accidently placed in a glass cell instead of a silica cell.   L/S

(iii) The sample cell is accidently contaminated with propanone.   L/S

(iv) A tungsten source is used instead of a deuterium source.   L/S

(v) The pH of the reaction system is not adjusted to the optimum value.   L/S

SAQ 2.1

2.2   REAGENTS, COMPLEXATION TECHNIQUES AND SOLUTION CONDITIONS

Spectroscopic methods of analysis in the UV/visible region of the electromagnetic spectrum show high sensitivity when the material being analysed absorbs strongly somewhere in the wavelength range between 200 and 800 nm. Sometimes the component being analysed has its own characteristically strong absorption, but more often it may require the addition of a special reagent to react selectively with the desired component to produce a derivative with the necessary high absorptivity.

∏   Do you recall what we mean by high absorptivity?

Let us return to the absorption of copper(II) sulfate in water referred to in Section 1.1. You should recall that a solution of $CuSO_4$ (0.01 mol dm^{-3}) had an absorbance of about 0.2 at 800 nm, which then increased to 0.8 at 600 nm when ammonia was added. The blue colour of the cuprammonium ion shows a reasonably linear Beer's law plot and has been used for the analysis of samples containing 10 mg or more of copper. However, other metal ions can also react with ammonia to produce blue coloured complexes, and a more selective colorimetric reagent for copper is diethyldithiocarbamate (DEDC), which is usually used as the sodium salt. This reagent produces a characteristic yellow-brown copper complex, having a maximum absorption at 436 nm, which can be readily extracted into organic solvents such as trichloromethane or carbon tetrachloride. It gives absorbance values of greater than 1.0 with microgram quantities of copper, i.e. at concentrations of about 10^{-4} mol dm^{-3}. Its absorptivity must, therefore, be high.

∏ By using the values and concentrations quoted above calculate approximate values for the molar absorptivity (ϵ_{max}) for Cu(II), Cu(NH$_3$)$_4^{2+}$ and the Cu–DEDC complex, assuming that the measurements were taken using 1 cm pathlength cells.

We rearrange Beer's Law to calculate ϵ from $\epsilon = A/cl$, and the following results are obtained:

	Cu(II)	Cu(NH$_3$)$_4^{2+}$	Cu–DEDC
λ_{max} (nm)	800	600	436
ϵ_{max} (dm^3 mol^{-1} cm^{-1})	20	80	10 000
ϵ_{max} (m^2 mol^{-1})	2	8	1 000

The reaction of copper ions with DEDC is simple and rapid, and the coloured complex is stable for more than an hour. By suitable adjustment of the solution conditions the reaction is highly selective for Cu(II). Furthermore, since aqueous solutions of the complex can be extracted readily into small volumes of organic solvents, minute traces of copper can be determined.

Therefore, for spectroscopic analysis it is important to choose the best available reagent and solution conditions for the analyte which is being determined.

2.2.1 The Ideal Reagent

Reagents for UV/visible spectroscopic analysis can vary from the metal chelating reagents, such as 1,10-phenanthroline for iron and

DEDC for copper, to enzyme reagents for the analysis of organic and biochemical species. Many enzymatic methods are based on recording the change of UV absorption of the reduced forms of nicotinamide–adenine dinucleotide (NADH) or the corresponding phosphate (NADPH). Although there is no such thing as a 'perfect' reagent, it is useful to judge the performance of a given reagent against the ideal properties listed below:

(a) *Stability in solution.* Some reagents deteriorate in a few hours while others ferment or grow moulds on storage. The instability of a reagent necessitates the use of freshly prepared solutions and recalibration of the spectrometer for each new batch.

(b) *Rapid and reproducible reaction.* For those analytical procedures which require the reaction to go to completion prior to the measurement being made, it is desirable that the reaction be rapid and reproducible. If the reaction is stoichiometric and the product and reagent absorb at different wavelengths, then it is often satisfactory and very convenient to use an excess of the reagent.

The stability of the product is also important and particularly so if measurements are not always made at the same time following initiation of the reaction. The stability may be temperature and/or time dependent.

(c) *Reproducible rate of reaction.* Many analytical procedures do not require the reaction to go to completion. Examples are those which employ automated methods in which timing can be very accurately controlled, and those such as enzymatic reactions in which the rate of reaction is measured. In these cases temperature control is often very important.

(d) *Selectivity or specificity of the reagent–analyte reaction.* This property is important in order that the absorption measured is that for the desired constituent only. This is sometimes difficult to achieve for complexing agents which are used for inorganic species. For example, 8-hydroxyquinoline will complex with an extensive range of metal ions under various pH conditions, and similarly, 2,4-dinitrophenylhydrazine forms derivatives with many aldehydes and ketones. These substances cannot be considered to be specific, but they can be used for the determination of a particular *class* of chemical species. In contrast, many biochemical reagents, particularly enzymes and immunoassay reagents based on monoclonal preparations, are highly selective.

(e) *Solvent compatibility*. The reagent should, of course, be soluble in the same solvent used for dissolving the sample and should react to give a complex which is still soluble over a wide concentration range.

(f) *Linear calibration*. The product of the reaction should obey Beer's law over a wide range of concentrations. This results in linear calibration graphs.

2.2.2 Choice of Reagent

For any given analyte there are probably a number of alternative reagents which could be used in spectroscopic analysis. For example, earlier in this chapter we suggested that sodium diethyldithiocarbamate was a better reagent for copper than ammonia, mainly because of its higher sensitivity and better selectivity. However, the blue copper–ammonia complex may be quite suitable for determining the copper content of certain steels, in cases where the concentration of copper is appropriate, the sensitivity requirements are not high and other reactive elements such as nickel are absent, or present only in low concentrations. In this situation the copper–ammonia method is rapid, inexpensive and avoids an organic extraction.

There are many reagents available for the analysis of iron, ranging from the thiocyanate method for iron(III) to a variety of reagents for iron(II). The thiocyanate method depends on the formation of a red colour in acid solution and the simplicity of the analytical procedure is one of its main attractions. The red complex which is formed, however, is non-stoichiometric and the colour is unstable. This instability is influenced by the concentration of the reagents, the ionic strength of the solution, and interference due to the presence of ions such as chloride and sulfate. The colour is also a function of pH.

Low concentrations of iron are, therefore, usually determined as iron(II) because a large number of nitrogen-containing organic reagents form stable and highly coloured complexes with this species. This generally involves an initial reduction step in which iron(III) is reduced to iron(II).

Some of the best known reagents for the determination of iron(II) are indicated below:

Reagent	ϵ_{max} $(\mathrm{dm^3\,mol^{-1}\,cm^{-1}})$	λ_{max} (nm)
2,2'-Dipyridyl	8 000	522
1,10-Phenanthroline	11 000	510
4,7-Diphenyl-10-phenanthroline	22 400	533
2,4,6-Tris(2-pyridyl)-1,3,5-triazine	22 600	595

The last reagent in this list (TPTZ) was selected by analysts in the water industry as the best available reagent for the determination of low concentrations of iron. In the report of their investigations they highlighted what they considered were the important features for suitable complexes as follows:

(a) The molar absorptivity (ϵ_{max}) of the complex should be of the order of 19 000 $\mathrm{dm^3\,mol^{-1}\,cm^{-1}}$ or more to ensure that the required analytical precision can be achieved, given the worst precision to be expected of the absorbance measurements.

(b) The coloured iron complex must be soluble and stable in aqueous solution so that solvent extraction procedures are not required.

(c) The chromogenic reagent should be commercially available and preferably of a similar cost to reagents in common use.

(d) The chromogenic reagent must be reasonably selective for iron, and should preferably react with Fe(II) rather than Fe(III).

The TPTZ reagent best fulfilled these requirements and was chosen as the preferred colorimetric reagent for iron analysis in the UK. The detailed procedure developed for its use is described in Chapter 3 of this Unit. In the USA, however, the prescribed reagent is still 1,10-phenanthroline.

2.2.3 Solution Conditions for Analysis

In the development of an analytical method for the analysis of a species in solution by spectroscopy, it is usual to check the accuracy, precision and detection limits of the proposed method with possible variations of solution conditions. Such variations should include the following:

(a) solvent polarity;

(b) pH and ionic strength (aqueous solutions);

(c) temperature.

Other factors which might well influence the analysis include:

(d) order of addition of reagents;

(e) mixing or stirring rate;

(f) time allowed for colour development.

The best methods should be independent of most of these variations, but ideal conditions rarely apply and a compromise is often necessary.

SAQ 2.2

Potassium thiocyanate and 1,10-phenanthroline have both been used as reagents for the determination of low concentrations of iron. Both have advantages and disadvantages for this application. Assign as many as possible of the following advantages and disadvantages to the two reagents.

Advantages

(i) Complex formation requires only the addition of the reagent and some acid.

(ii) The reagent is cheap.

(iii) The complex is stable and relatively free from interferences.

→

SAQ 2.2
(Contd)

(iv) The molar absorptivity is over $1000\,\mathrm{m^2\,mol^{-1}}$.

(v)   It is applicable to iron in the Fe(III) state.

Disadvantages

(vi) The iron must be reduced to the Fe(II) state.

(vii) The complex is non-stoichiometric.

(viii) The molar absorptivity is too low to be chosen for water analysis (in the UK).

(ix) A control of pH is important.

(x)   The complex is sensitive to light, and is relatively unstable.

2.3 CHOICE OF WAVELENGTH AND CALIBRATION DATA

In the previous two sections we have looked at the factors which have to be considered when a sample is being prepared for spectroscopic analysis in the UV/visible region of the spectrum. In this section we will consider the optimum instrumental settings necessary to achieve the highest precision possible in the analysis. At this stage we will assume that we already have the sample in solution at the correct concentration level and that the solution is stable, and hence the absorbance is constant with time. We merely have to place the solution into the spectrophotometer, choose the appropriate wavelength for the measurement, take suitable measurements of the amount of radiation absorbed and deduce the concentrations by reference to a set of suitable calibration data. In order to illustrate the procedure we will go through the analysis of the manganese content of a steel. This procedure is based upon the oxidation of the manganese to the purple coloured permanganate (manganate(VII)) ion in aqueous solution. It is first necessary to prepare a set of calibrations solutions. We can use the set of solutions already used in Section 1.2.2.

2.3.1 Calibration Data Using Permanganate Solutions

The standard solutions for potassium permanganate prepared in Section 1.2.2 have the following manganese concentrations:

Solution	Blank	A	B	C	D	E
Mn concentration $(mg\,dm^{-3})$	0.00	5.00	10.0	15.0	20.0	30.0

Visible absorption spectra of these solutions were recorded over the wavelength range from 400 to 650 nm by using a double-beam spectrometer. These spectra are shown in Figure 2.3a. As expected, this series of solutions shows that the intensity of the purple colour increases as the manganese content increases.

Before recording these spectra the instrument was set at 520 nm and the zero absorbance (100% transmittance) checked. Two cells, which had been thoroughly washed, were filled with distilled water, and one was placed in the sample beam, with the other in the reference beam. The absorbance was then adjusted to zero. Only a slight adjustment was necessary, but if the cells had been perfectly optically matched at 520 nm no adjustment would have been needed. The zero was recorded and this represents the spectrum of the water blank used in this experiment. You will see that the blank deviates slightly from zero at about 500 nm, but is within 1 % T of 100 % T over the whole range.

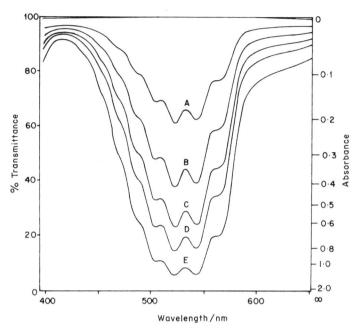

Figure 2.3a Visible absorption spectra of potassium permanganate solutions (as in Figure 1.2c)

∏ Complete the following table of data by making measurements at λ_{max} = 522 nm on the six spectra shown in Figure 2.3a, or use the measurements that you made from Figure 1.2c (remember that A = $\log(100/\% T)$).

λ		Blank	A	B	C	D	E
λ_{480}	% T	99.9	82.9	69.0	57.5	48.5	38.5
	A	0.000	0.082	0.161	0.240	0.314	0.475
λ_{max}	% T						
	A						

The absorbance data at λ_{max} should be similar to those reported below:

0.000	0.211	0.421	0.638	0.836	1.236

∏ Plot the absorbance for λ_{480} and λ_{max} against the concentration of manganese and draw the best straight-line calibration graphs. Is Beer's law obeyed in both cases (i.e. at 480 and 522 nm)?

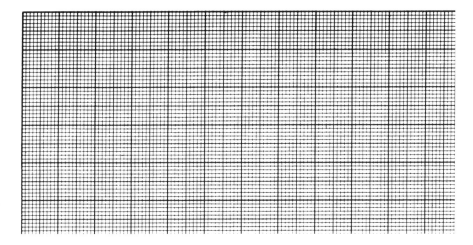

Both calibration graphs are linear over the concentration range from 0 to 30 mg dm^{-3}, and pass through the origin (see Figure 2.3b). Hence Beer's law is obeyed at both wavelengths over the full concentration range being employed. An additional data point at 25 mg dm^{-3} would have helped in drawing the best line, particularly at the high-

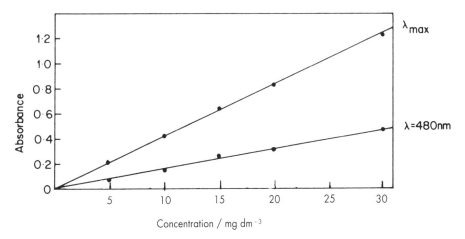

Figure 2.3b Calibration curves for the determination of Mn at 480 and 522 nm

concentration end. You will probably agree that it is more difficult to make measurements at 480 than at 522 nm because in the former case we are taking measurements from the side of a peak, which thus limits the accuracy. Should a small error occur in the reading of the wavelength scale, this will result in a considerable error in the corresponding reading taken from the % T or A scale. In addition, we saw in the previous chapter that we are more likely to get deviations from Beer's law because of using polychromatic radiation when measurements are not made at λ_{max}. The calibration graph for 480 nm shows a slightly greater scatter about the calibration line, and being of lesser slope results in reduced accuracy and sensitivity of the analysis. Whenever possible measurements should be taken at λ_{max} or at the top of an alternative absorption peak.

SAQ 2.3a

Calculate the absorptivities of $KMnO_4$ using the following data:

A $KMnO_4$ solution at λ_{max} = 522 nm gave an absorbance of 1.236 in a 10 mm cell.

The Mn concentration is 30 mg dm^{-3} ($A_r(Mn)$ = 54.938).

(i) Molar absorptivity, ϵ_{max};

(ii) Absorptivity, $E_{1\%}^{1\,cm}$.

SAQ 2.3a

2.3.2   The Determination of Manganese in Steel

1.000 g of steel was dissolved in acid, the manganese oxidised to permanganate, and the volume of the solution made up to exactly 100 cm^3. The visible absorption spectrum of this solution is shown in Figure 2.3c.

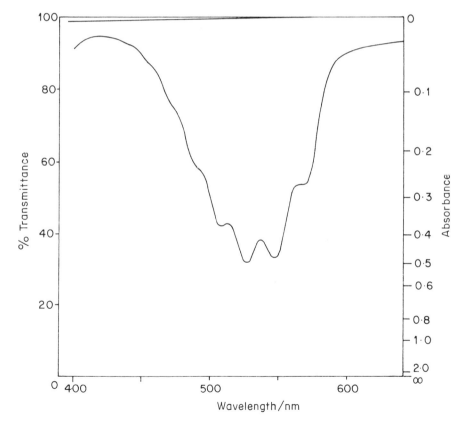

Figure 2.3c   Visible absorption spectrum of a solution of potassium permanganate of unknown concentration

∏   Read off the transmittance of the steel solution at λ_{max} and then calculate the corresponding absorbance.

You should have obtained a transmittance of 33%, giving an absorbance of 0.5.

∏   Now read off the concentration of manganese in the steel solution using the permanganate calibration curve you produced by using the data for the permanganate standards.

Your result for the concentration of Mn in the steel should be similar to my value of $12\,mg\,dm^{-3}$.

The concentration of steel in this solution is 1 g per $100\,cm^3$, i.e. $10\,g$ dm^{-3}. Therefore, the amount (in g) of Mn in $100\,g$ of steel (i.e. wt%) is 120×10^{-3}. The Mn content of the steel is therefore $0.12\,wt\%$.

Throughout this exercise we have used a recording UV/visible spectrometer with an ordinate scale which was linear in transmittance. This has enabled us to consider a number of aspects of analysis such as the following:

(a) calculation of absorbance from transmittance values;

(b) selection of the analysis wavelength;

(c) calibration curves at λ_{max} and other wavelengths;

(d) potential errors with respect to quantitative spectrometric measurement.

We could, of course, have carried out this determination of Mn in steel by using a number of other types of instrument, such as a simple manual colorimeter operating in the visible region only and having a digital display which reads out directly in absorbance values.

2.3.3   The Sample Blank

There is an important aspect of the analysis that we have so far ignored, but which can seriously affect accuracy if it is not taken into account. The standards and sample solutions were of quite different composition and were produced as a result of quite different chemical procedures.

The standard solutions were made up by taking a high-purity sample of $KMnO_4$ and dissolving it in deionised water. The steel sample, however, was dissolved in acid, subjected to oxidation and finally diluted with deionised water. The most obvious difference in the composition is the presence of a range of elements in the solution of the steel, and in particular the presence of a very large concentration of iron.

∏ Is it necessary to exactly match the composition of the standards with the sample?

Not always, but we need to find out before publishing our methods and results.

One of the simple ways of testing for accuracy is to purchase an analysed standard sample which has a composition similar to your own sample. The standard can then be analysed and the value(s) checked against the certificate values provided with the standard.

This approach is fine if such a standard is available, but this is often not the case. What should we do then?

You could employ an alternative method of analysis which is quite different to the one of interest. If the results are in close agreement then you can be quite confident in your accuracy, because it would be very unusual for the errors associated with two different methods to be the same.

The evaluation of accuracy is a big subject in its own right and you need to be aware of it in all your analyses. In Section 2.4 we will describe standard addition methods which are alternative approaches to the problems associated with these so-called matrix effects.

Let us return to our steel analysis. From where might the errors arise?

(a) The reagents used (acid and oxidising reagent) may themselves contain impurities, including manganese itself, which can give rise to absorption at, or very close to, the analysis wavelength of 522 nm.

(b) The constituents of steel, other than the manganese, may give rise to coloured products which absorb at the analysis wavelength.

Because this mismatch between the sample and standards is so great in terms of the iron content, we could test for iron interference. This can be done by obtaining very high purity iron wire (99.99%) and subjecting it to the dissolution, oxidation and solution procedures. If you do this a yellow solution is obtained which has zero absorbance at 522 nm. That clears the oxidising agent and iron, but leaves possible problems due to other constituents of the sample.

We will not pursue this any further at this stage, but you should be on your guard if your results are to be accurate. We have looked at one specific example but the lessons it has taught us are equally applicable to any other sample you may encounter, such as river water, blood plasma, sausage meat, etc.

2.3.4   Wavelength and Absorbance Checks

Most visible spectrometers are supplied with a set of standard filters to be used for checking wavelength calibration over the visible region. An alternative method which can be employed is to use the atomic emission lines from a suitable vapour discharge lamp. The filters commonly available are rare earth (holmium and didymium) oxide glasses which give a series of sharp absorption bands. Some typical wavelength values are shown in Table 2.3, below.

Considerable variations occur, however, between different batches of glass, and they are now considered to be unreliable for the calibration of high quality instruments. The most accurate method of wavelength calibration is by introducing a discharge lamp into the lamp housing of

Table 2.3 Wavelengths (nm) of rare earth absorption filters used for wavelength calibration

Holmium glass	Didymium[a] glass
241.5 ± 0.2	
279.4 ± 0.3	573.0 ± 3.0
287.5 ± 0.4	586.0 ± 3.0
333.7 ± 0.6	685.0 ± 4.5
360.9 ± 0.8	
418.4 ± 1.1	
453.2 ± 1.4	
536.2 ± 2.3	
637.5 ± 3.8	

[a]Neodymium and praseodymium

the spectrometer. The most commonly used discharge lamps for this purpose are those of neon and mercury. Some typical wavelength values are as follows:

Neon emission (nm)	Mercury emission (nm)
533.1	–
534.1	253.7
540.1	364.9
585.3	404.5
594.5	435.8
614.3	546.1
633.4	576.9
640.2	579.0

Apart from these special sources, deuterium emission lines are particularly useful for a quick check, since most UV instruments have a deuterium lamp as a source. The lines used are the red line at 656.1 nm and the blue–green line at 486.0 nm.

An appropriate method, which may be used in the visible region with an instrument fitted with a tungsten source, is to observe the position (by using a small mirror) where the light passing through the sampling compartment appears to be pure yellow in colour, i.e. neither reddish nor greenish. For most observers with normal colour vision this is observed between 570 and 580 nm.

The absorbance scale of an instrument is best checked by using one of the recommended solution standards listed below. The use of solution standards combines both operator error and cell errors in one set of measurements and, providing that Beer's law is obeyed, can be used over any absorbance range within the instrument's capability. The disadvantage of solution standards is that they require careful preparation and show some temperature dependence. Solutions which have been used as standards include the following:

(a) potassium dichromate in acid solution;

(b) potassium dichromate in alkaline solution;

(c) potassium nitrate;

(d) pyrene in isooctane;

(e) potassium hydrogen phthalate;

(f) nicotinic acid.

Details of the standard values and the wavelength ranges over which the standards can be used are listed in a the book, C. Burgess and A. Knowles, *Standards in Absorption Spectrometry* (Techniques in Visible and Ultraviolet Spectrometry: Vol. 1), Chapman and Hall, London, 1981.

SAQ 2.3b A solution of potassium permanganate with a manganese concentration of 3.4 mg dm^{-3} transmits 23% at 522 nm and 57.5% at 480 nm. Calculate the effect of a +1% transmittance error on the absorbance at these two wavelengths. Indicate, giving reasons, which wavelength it is best to use for the analysis of permanganate solutions (522 nm is the λ_{max} for KMnO$_4$; 480 nm is on the side of the adsorption band).

2.4 STANDARD ADDITION METHODS

2.4.1 Matrix Effects in Spectrophotometric Analysis

In Section 2.3 we discussed the factors which may affect the accuracy of a spectrophotometric analysis, and one of the factors which was identified was the need, ideally, for the composition of the standards to match roughly the composition of the sample to be analysed, not only with respect to the concentration of the analyte but also with regard to the concentrations of other species in the sample. We saw in the determination of the manganese content in steel that the composition of the sample solution was very different from that of the standards and we needed to be aware of these differences when carrying out the analysis. In this case, we knew that iron would be present in a very large excess and so we could test for any possible interference by iron in our determination of manganese.

Matrix effects may, however, be important because, for example, the absorbance of many coloured complexes of metal ions is decreased by varying amounts of sulfate or phosphate ions as a result of the tendency of these anions to form colourless complexes with metal ions. As a result, the colour-forming reaction is incomplete and so the measured absorbance is lowered. The effects of sulfate and phosphate can often be countered by introducing into the standards amounts of these anions which approximate to the amounts found in the sample.

Often, however, we may not know what other species are present in a sample, for example when complex materials such as minerals, soils and plant ash are analysed, and it is therefore impossible to make up standard solutions which have a similar composition to our sample. When this is the case the method of *standard addition* is often useful in counteracting matrix effects.

2.4.2 Principles of Standard Addition

The standard addition method can take several forms. The method that will be described here involves adding one or more increments of a standard solution to sample aliquots of the same size. Each solution is then diluted to the same volume before the absorbance is measured. If you have previously completed the Units on Polarography and Voltammetric Methods or Potentiometry and Ion Selective Electrodes in the ACOL Series, you may already be familiar with a modified version of standard addition in which successive increments of a standard solution are added to a single, known amount of the unknown. Measurements are then made on the original unknown solution and after each addition of the standard. This is particularly useful if the amount of sample is limited.

We will assume, initially, that we have a large volume of our unknown solution. It is now necessary to define a set of symbols so that we can derive an expression in order to calculate the concentration of our analyte in the sample solution.

Let:

V_x = volume of the unknown solution transferred to the volumetric flask in each case;

C_x = concentration of the analyte in the unknown solution;

V_t = volume of the volumetric flask;

V_s = variable volume of the standard solution of the added analyte;

C_s = concentration of the standard solution of the analyte.

We will also need to add our colour-forming reagents before the solution is made up to the mark in the volumetric flask.

If Beer's law is obeyed, the total absorbance measured at λ_{max} for the analyte complex will be as follows:

$$A_{total} = A_{sample} + A_{standard} \tag{2.1}$$

where A_{sample} is the absorbance due to the amount of analyte in the sample solution (which will be the same for all of the solutions), and $A_{standard}$ is the absorbance due to the amount of the analyte added in the standard solution.

We can rewrite equation (2.1), by using the parameters that we defined previously, as follows:

$$A_{total} = \frac{\epsilon l V_x C_x}{V_t} + \frac{\epsilon l V_s C_s}{V_t} \tag{2.2}$$

In equation (2.2), ϵ, l, V_x, C_x, V_t and C_s are all constants, and therefore A_{total} will depend only on the volume V_s. Equation (2.2) can therefore be written as follows:

$$A_{total} = \alpha + \beta V_s \tag{2.3}$$

where

$$\alpha = \frac{\epsilon l V_x C_x}{V_t} \quad \text{and} \quad \beta = \frac{\epsilon l C_s}{V_t}$$

You should recognise that equation (2.3) is the equation for a straight line in which α is the intercept on the y-axis and β is the gradient.

Therefore, in order to calculate the unknown concentration C_x, we plot a graph of A_{total} against the volume of standard added, V_s. We can use a least-squares analysis to determine the gradient β and the intercept α. C_x can then be obtained from the ratio of these two quantities and the known values of V_x and C_s, since the following applies:

$$\frac{\alpha}{\beta} = \frac{\epsilon l V_x C_x / V_t}{\epsilon l C_s / V_t} = \frac{V_x C_x}{C_s}$$

or

$$C_x = \frac{\alpha C_s}{\beta V_x} \tag{2.4}$$

2.4.3   The Determination of Iron in Drinking Water

In this method 20.00 cm³ aliquots of a water sample were pipetted into 100 cm³ volumetric flasks. Exactly (0.00), 5.00, 10.00, 15.00, or 20.00 cm³ of a standard solution containing 15.0 mg dm⁻³ of Fe(III) were added to each, followed by an excess of ammonium thiocyanate solution in order to form the characteristic red complex, $Fe(CNS)^{2+}$. After making up to the mark with distilled water the absorbance of the five solutions were measured by using water as the blank.

Volume (cm³)	0.00	5.00	10.00	15.00	20.00
Absorbance	0.215	0.424	0.621	0.826	1.020

Π Plot a graph of absorbance against volume of standard added and draw the best straight line. Calculate the gradient and the intercept on the *y*-axis.

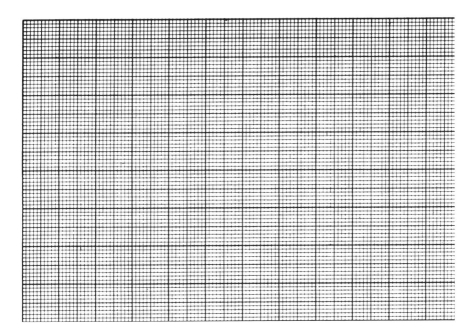

The graph is linear over the range of volumes of standard used (see Figure 2.4). I calculated a gradient of 0.0401 and an intercept of 0.2250. Did your results agree? We can now calculate the concentration of iron in the water by using equation (2.4).

We know that:

$V_x = 20.00 \, \text{cm}^3$

$C_s = 15.00 \, \text{mg dm}^{-3}$

$\alpha = 0.2250$

$\beta = 0.0401$

Therefore, by using equation (2.4) we find the following:

$$C_x = \frac{0.2250 \times 15.00}{0.0401 \times 20.00} = 4.21 \, \text{mg dm}^{-3} \, \text{Fe(III)}$$

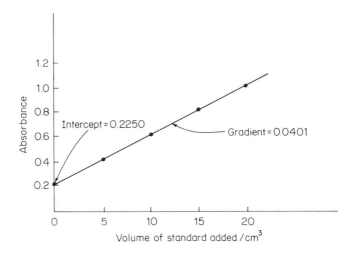

Figure 2.4   Standard addition plot

2.4.4   A Simplified Method for Standard Addition

The standard addition method described so far requires a large amount of sample. In our example we would have needed at least 100 cm^3 of water in order to carry out the analysis. Where we only have a limited amount of sample it is possible to perform a standard addition analysis by using only two increments of sample. In this method a single addition of a known volume of standard is added to one of the two sample volumes.

If Beer's law is obeyed we can write the following:

$$A_1 = \frac{\epsilon l V_x C_x}{V_t} \tag{2.5}$$

and

$$A_2 = \frac{\epsilon l V_x C_x}{V_t} + \frac{\epsilon l V_s C_s}{V_t} \tag{2.6}$$

where A_1 and A_2 are the absorbances of the diluted sample and the diluted sample plus standard, respectively.

By rearrangement of these two equations we can obtain an expression for the unknown concentration C_x, namely:

$$C_x = \frac{A_1 C_s V_s}{(A_2 - A_1) V_x} \qquad (2.7)$$

SAQ 2.4

A $10.0 \, \text{cm}^3$ aliquot of an aqueous solution of quinine was diluted to $25 \, \text{cm}^3$ and found to have an absorbance of 0.217 at 348 nm when measured in a 1.00 cm pathlength cell. A second $10.0 \, \text{cm}^3$ aliquot was mixed with $5.00 \, \text{cm}^3$ of a solution containing 27.3 parts per million (ppm) of quinine. After dilution to $25.0 \, \text{cm}^3$ this solution had an absorbance of 0.474 when measured in the same 1.00 cm pathlength cell. Calculate the amount of quinine, in ppm, in the original aqueous solution.

2.5 PHOTOMETRIC TITRATIONS

Photometric or spectrophotometric measurements can be used to advantage in locating the end-point in a titration, provided that the analyte, the reagent or the titration product absorbs electromagnetic radiation. Alternatively, an absorbance indicator can provide the absorbance change necessary to locate the end-point.

2.5.1 Titration Curves

A photometric titration curve is a plot of absorbance, corrected for changes of volume during the titration, as a function of the volume of titrant. If conditions are chosen properly, the curve will consist of two straight regions, having different slopes, which intersect at the end-

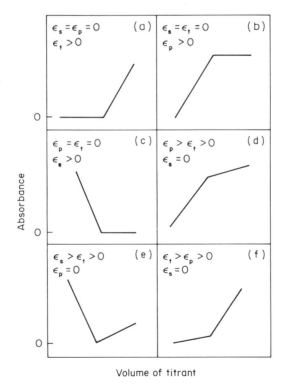

Figure 2.5a–f Typical photometric titration curves, where the absorptivities for sample titrated, product and titrant are given by ϵ_s, ϵ_p and ϵ_t, respectively

point. The shapes which are obtained depend on whether it is the analyte, the titration reagent or the product which is coloured. Titration of a non-absorbing species with a coloured titrant that is decolourised by the reaction produces a horizontal line in the initial stages of the titration, followed by a rapid rise in the absorbance beyond the end-point (see Figure 2.5a). The formation of a coloured product, from colourless reactants, initially produces a linear rise in the absorbance, followed by a region in which the absorbance becomes independent of the volume of titrant added (Figure 2.5b). Depending upon the absorption characteristics of the reactants and products, the other curves shown in Figures 2.5c–f can also be obtained.

In order to obtain a satisfactory photometric end-point it is necessary that the absorbing species obey Beer's law, otherwise the titration curve will not have the linear regions necessary for extrapolation to the end-point. It is also necessary to correct the measured absorbance for volume changes. If the original volume of the analyte solution is V cm^3 and we add v cm^3 of titrant we need to multiply the observed absorbance by $(V+v)/V$ in order to obtain the corrected absorbance.

2.5.2 Instrumentation

Photometric titrations are usually performed with a spectrophotometer that has been modified to permit insertion of the titration vessel into the lightpath. After the zero adjustment of the absorbance scale (using a suitable blank solution) has been made, light is allowed to pass through the solution of the analyte and the instrument is adjusted until a convenient absorbance reading is obtained. Usually no attempt is made to measure the true absorbance since relative values are adequate for the purposes of end-point detection. Data for the titration are then collected without varying the instrument conditions.

The power of the radiation source and the response of the detector should remain constant throughout the titration. Cylindrical containers are usually used for the titration and care must be taken to avoid moving the container during the procedure since this may affect the pathlength.

2.5.3   Determination of Bismuth and Copper in a Mixture

The photometric end-point has been used a great deal in titrations with ethylenediaminetetraacetic acid (EDTA) and other complexing reagents. EDTA reacts with a large number of metal cations to form complexes. At 745 nm, neither bismuth(III), copper(II) nor the EDTA reagent absorbs, nor does the more stable bismuth complex, which is formed in the first part of the titration. The copper complex with EDTA does, however, absorb at this wavelength.

∏   Sketch the photometric titration curve that you would expect to see when a mixture of Bi(III) and Cu(II) is titrated with EDTA solution.

Your titration curve should look similar to that shown in Figure 2.5g. The solution shows no absorbance until all of the Bi(III) has been titrated. With the first formation of the copper(II) complex an increase in absorbance occurs until all the copper(II) has reacted. Further reagent additions cause no further absorbance change and the two end-points are thus clearly defined.

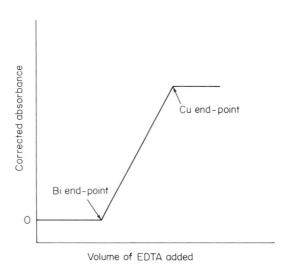

Figure 2.5g   Photometric titration curve for a mixture of Bi and Cu titrated with EDTA, where the values for the absorbance have been corrected for volume changes

2.5.4 Applications of Photometric Titrations

Photometric titrations often give more accurate results than a direct photometric analysis because the data from several measurements are pooled to give the final end-point. Furthermore, the presence of interfering species in the solution may not affect the result, since only a change in absorbance is being measured. For example, it has been shown that considerable amounts of zinc, cadmium, tin(IV), manganese(II), and chromium(III), and smaller amounts of aluminium(III), cause little or no interference in the photometric titration of iron(III) with EDTA at pH 2.4.

The photometric end-point has the advantage over many commonly used end-points in that the experimental data are collected well away from the equivalence point and so the titration reactions need not have such favourable equilibrium constants as those required for a titration that depends upon observations near the equivalence point. For this same reason more dilute solutions can therefore be titrated.

The photometric end-point has been applied to all types of reactions. Most of the reagents used for oxidation–reduction titrations have characteristic absorption spectra, and thus produce photometrically detectable end-points. Photometric titrations are not restricted to inorganic species, however. Phenols can be titrated with tetra-n-butylammonium hydroxide by using propan-2-ol as the solvent. The absorbance measurements are made at the λ_{max} for the phenol in the ultraviolet region of the spectrum and in this way substituted phenols can be differentiated.

Summary

Quantitative determinations in UV/visible spectroscopy are dependent upon reliable methods of sampling materials, the treatment required to get all the components into solution and any chemical reaction used to obtain a coloured or highly UV-active species. The UV/visible absorption for the sample material has to be compared with a calibration graph previously prepared from a series of standard solutions of known concentrations in order to establish the concentration of the analyte in the sample.

Objectives

On completion of this chapter, you should now:

● understand the importance of the various steps involved in sampling and in preparing a particular sample for analysis;

● have acquired a knowledge relating to solvents and the effects of acidity which may determine the quality of the absorption spectra obtained from samples in solution;

● appreciate the important features in the choice of a reagent in producing a stable derivative with a strong absorption peak;

● be able to select the most suitable wavelength from a spectrum for use in quantitative measurements;

● have learnt the importance of wavelength and absorbance checks on spectrometers.

3. Spectroscopic Determinations

In Chapters 1 and 2 we considered the hardware which is used for spectroscopic analysis in the UV/visible region, i.e. the solvents and cells, the reagents and solution additives, the instruments and their characteristics, and the factors which limit the precision of methods of spectrophotometric analysis. In this chapter we consider the procedures used in analysis, the manner in which we can manipulate the data, and the range of applications which are available to us.

3.1 TYPICAL ANALYTICAL PROCEDURES

Let us return to the question posed at the beginning of Chapter 2, i.e. how would you go about determining the iron content of water using a colorimetric method and a visible spectrometer?

∏ Can you now list the five or more steps which need to be specified? You can — good! If not, then refer back to that section and remind yourself.

But how much detail can or should you remember?

Obviously an analytical technician engaged in repetitive and routine analysis will quickly learn the precise details of amounts to weigh, volumes of reagent solutions to add and the sampling and operational conditions for the instrument being used. However, it is not the intention of this course to supply these details, but instead we hope that you will learn to appreciate the items of detail you need to know and, most importantly, where you can find the necessary information.

In your working environment you will be aware of the sources of information and analytical procedures available immediately to you

— but what about information available elsewhere? There are, of course, standard methods laid down by the British Standards Institute (BSI) and the American Society for Testing of Materials (ASTM), as well as methods specified within a particular industry (e.g. the food or water industries). Whatever your sources, they need to provide you with all the information you require in order to carry out the analysis with the optimum accuracy and precision. However, not everything can be learnt from books or the analytical literature — you need to have some 'hands-on' experience of the particular analysis being attempted before you can expect to achieve reliable results in which other people will place confidence.

In this chapter we will start by looking at contrasting methods for the analysis of single components, such as iron and glucose. After this we shall consider the details of an investigation into the development of a method for ultra-trace-level analysis of iron. Next, we will look at the principles involved in the analysis of mixtures, which lead to the question of the manipulation of spectrophotometric data. Finally, we will summarise the application range of spectrophotometric methods covering the ultraviolet and visible region.

3.1.1 Spectrophotometic Methods for Iron in Water

Standard methods for the analysis of drinking waters are defined by government bodies in most countries. In the UK one of the joint technical committees of the Department of the Environment and the National Water Council is responsible for the provision and publication of recommended methods for the analysis of water. However, because both needs and the availability of equipment vary widely, a selection of methods may be recommended for a single component. In the USA the analytical procedures are given in the publication *Standard Methods for the Examination of Water and Waste Water*, which is published jointly by the American Public Health Association, the American Waterworks Association and the Water Pollution Control Federation.

In the latest edition of this publication two methods are listed for the determination of iron in water. Of the two methods one uses a procedure which is known as atomic absorption. This is said to be relatively easy to carry out, with both the precision and accuracy being superior to those of the second method, which uses a colorimeter. However, the colorimetric method involving 1,10-phenanthroline requires much less expensive instrumentation and is simple and reliable, and is still used extensively. In the US publication the determination of iron content by 1,10-phenanthroline is detailed under six headings, as follows:

1. General Discussion (principles, interference, concentration levels).

2. Apparatus.

3. Reagents.

4. Procedure.

5. Calculation.

6. Precision and Accuracy.

The corresponding standard method in the UK, *Iron in Raw and Potable Waters by Spectrophotometry (1977)*, published by Her Majesty's Stationery Office (HMSO), specifies the use of 2,4,6-tripyridyl-1,3,5-triazine (TPTZ) as the preferred colorimetric reagent for iron. This publication details the method of analysis under 14 headings, as follows:

1. Performance Characteristics of the Method.

2. Principle.

3. Interferences.

4. Hazards.

5. Reagents.

6. Apparatus.

7. Sample Collection and Preservation.

8. Sample Pretreatment.

9. Analytical Procedure.

10. Measurement of Absorbance.

11. Preparation of Calibration Curve.

12. Change in Concentration Range of the Method.

13. Checking the Accuracy of Analytical Results.

14. References.

As this Unit is concerned with the practical aspects of analysis itself, it is assumed that sample collection and preservation, and sample pretreatment (Sections 7 and 8) have been correctly carried out. We are particularly interested in the analytical procedure, and Sections 9 and 10 are reproduced below. Read through them carefully and note that guidance is given on the following:

(a) temperature control (9.1);

(b) time factor (9.2);

(c) cell size and measurement wavelength (9.3);

(d) the importance of blank determinations (9.4 and 9.7);

(e) compensation for sample colour and turbidity (9.6);

(f) calculation of the results (9.9–9.11).

With all methods of analysis you should be aware of the scope of the method as well as sources of error and the precision and accuracy. These are specified in Section 1 of the HMSO publication. They show that the reliability of the procedure has been evaluated in several laboratories in order to clearly establish its limitations.

Iron in Raw and Potable Waters by Spectrophotometry (1977 version)

Note: Throughout this method iron is expressed as the element (Fe)

1 Performance Characteristics of the Method

(For further information the determination and definition of performance characteristics see another publication in this series).

1.1	Substance determined	All forms of iron (see Sections 2 and 8).
1.2	Type of sample	Raw and potable waters.
1.3	Basis of method	Reduction of iron to the ferrous state and subsequent reaction with 2, 4, 6-tripyridyl-1, 3, 5-triazine to form a coloured complex whose concentration is measured spectrophotometrically.
1.4	Range of application (a)	Up to 1 mg/l.
1.5	Calibration curve (a)	Linear to 2 mg/l at 595 nm.

1.6 Total standard deviation

		Iron concentration (mg/l)	Total standard deviation (mg/l)		Degrees of freedom
1.6.1	Without pretreatment:	0.150	0.005	(a) (d)	15
		0.250	0.004	(a) (c)	17
		0.400	0.003–0.008	(b) (c)	6–9
		1.000	0.011	(a) (c)	19
1.6.2	With pretreatment:	0.250	0.013	(a) (c)	13
		0.334	0.019	(a) (d)	8
		1.000	0.013	(a) (c)	13

1.7	Limit of detection	
	1.7.1 Without pretreatment (b)	0.003–0.015 mg/l (with 5 to 10 degrees of freedom)
	1.7.2 With pretreatment (a)	0.06 mg/l (with 7 degrees of freedom).
1.8	Sensitivity (a)	1.0 mg/l gives an absorbance of approximately 1.25
1.9	Bias (a)	No important sources of bias were detected.
1.10	Interferences (a)	None of the substances tested caused appreciable errors except commercial polyphosphate (See Section 3).
1.11	Time required for analysis (a)	The total analytical and operator times are the same. Typical times for 1 and 10 samples are approximately 45 and 60 minutes excluding any pretreatment time.

(a) These data were obtained by the Water Research Centre (Medmenham Laboratory)[1] using this method and a spectrophotometer with 40-mm cells at 595 nm.
(b) These data were obtained from an interlaboratory calibration exercise in which 5 laboratories took part.[2]
(c) These data were obtained using distilled water spiked with the stated iron concentration.
(d) River Thames water.

9 Analytical Procedure

READ SECTION 4 ON HAZARDS BEFORE STARTING THIS PROCEDURE

Step	Experimental Procedure	Notes

Analysis of samples

9.1 Add 40.0±0.5 ml of the well mixed sample to a 50-ml calibrated flask. Adjust the temperature of the sample, if necessary, to between 15 and 30 °C (notes e and f).

(e) If the sample contains polyphosphate, see Section 3 note (g).
(f) See Section 12 for concentration range.

9.2 Add to the flask, swirling after each addition, 2.0±0.1 ml of 10% m/V hydroxylammonium chloride solution, 2.0±0.1 ml of 0.075% m/V TPTZ solution, and 5.0±0.2 ml of acetate buffer solution. Dilute with water to the mark, stopper the flask, and mix the contents well (notes g and h). Allow to stand between 5 minutes and 2 hours.

(g) If a batch of samples is to be analysed, each reagent can be added to all samples before adding the next reagent.
(h) If pretreatment has been used see Section 8.1.

9.3 Meanwhile set up the spectrophotometer (see Section 6.2) according to the manufacturer's instructions. Adjust the zero of the instrument with water in the reference cell. Measure the absorbance (see Section 10) of the well mixed solution at 595 nm using 40-mm cells (note i). Recheck the instrument zero. Let the absorbance of the sample be S.

(i) Other sizes of cells may be used but the performance characteristics quoted in Section 1 would no longer apply.

Blank determination (if pretreatment not required)

9.4 A blank must be included with each batch (eg up to 10 samples) of determinations for which pretreatment was not required using the same batch of reagents as for samples. Add 0.80±0.05 ml of 5M hydrochloric acid and 39±1 ml of water to a 50-ml calibrated flask and adjust the temperature to between 15 and 30 °C.

9.5 Carry out steps 9.2 and 9.3, let the absorbance of the blank be B.

Step	Experimental Procedure	Notes

Compensation for colour and turbidity in the sample (note d)

9.6 A sample compensation solution must be included with each sample for which a colour/turbidity correction is necessary using the same batch of reagents as for samples. Carry out steps 9.1 to 9.3 but omitting addition of the TPTZ reagent. Let the absorbance of the sample compensation solution be S_1.

Determination of iron in the water used for the blank (notes j and k)

9.7 Add 1.60 ± 0.05 ml of 5M hydrochloric acid and 29 ± 1 ml of water to a 50-ml calibrated flask and adjust the temperature to between 15 and 30 °C. Add to the flask, swirling after each addition, 4.0 ± 0.1 ml of 10% m/V hydroxyammonium chloride solution, 4.0 ± 0.1 ml of 0.075% m/V TPTZ solution and 10.0 ± 0.2 ml of acetate buffer solution. Dilute with water to the mark, stopper the flask, mix the contents well and carry out step 9.3. Let the absorbance be D.

(j) This determination is not needed if the iron content of the water used for the blank is known or is negligible (Section 13.3).

(k) All reagents must be from the same batch as for the samples.

9.8 The absorbance due to iron in 50 ml of water W is given by:
$$W = 2B - D - C$$
where C = absorbance of sample cell when it and the reference cell are filled with water. Calculate the iron concentration in the water C_w from 0.8 W (note l) and the calibration curve. (See Section 11).

(l) The factor 0.8 allows for the fact that the calibration curve is for 40 ml samples whereas W was obtained for an effective 50 ml sample.

Calculation of results

9.9 Calculate the apparent absorbance due to iron in the sample, R, from
$$R = S - B$$
or, when a correction for colour/turbidity is made
$$R = S - B - S_1 + C$$

9.10 Determine the apparent iron concentration, C_a, in the sample from R and the calibration curve. (See Section 11).

9.11 Calculate the iron concentration in the original sample, C_r, from
$$C_r = 1.02\,(C_a + C_w) \text{ mg/l (note m)}$$

(m) The factor 1.02 allows for the dilution of the sample by the acid into which it was collected. (See Section 7).

10 Measurement of Absorbance

The exact instrument setting for the wavelength of the absorption peak must be checked for each instrument and then used for all future work. The procedure used for measuring absorbance should be rigorously controlled to ensure satisfactory precision. The same cells should always be used and should not be interchanged between the reference and sample. They should always be placed in the same position in the cell holder with the same face towards the light source.

It is difficult to ensure reproducible alignment of cells with chipped corners, and therefore they should be discarded. Similarly the slide of the cell holder should be kept scrupulously clean. Before every set of measurements the absorbance of the sample cell should be measured against the reference cell when both are filled with water. This will also enable the true absorbance of the blank to be determined.

SAQ 3.1a

In many colorimetric analyses it is sufficient to correct the sample absorbance (S) for the blank reading (B), and to read off the component concentration from the Beer's law calibration graph. This is implied in the above procedure for iron in water when the apparent absorbance is given by:

$$R = S - B$$

Indicate if when using the above simple procedure the factors below would result in a high (H), low (L) or correct (C) value for the iron content of the water being analysed. If you have insufficient information to make a judgement use the code (I).

(i) The temperature of the sample dropped to 15°C before measurement.

(ii) The instrument was found to have a 1 nm calibration error.

(iii) The sulfuric acid reagent was found to contain $1 \, mg \, dm^{-3}$ iron.

(iv) The deionised water used for the blank was found to contain $0.1 \, mg \, dm^{-3}$ iron.

(v) Only 1/10 of the quantity of hydroxylammonium chloride reagent was added ($0.2 \, cm^3$ instead of $2.0 \, cm^3$).

(vi) The final solution for measurement looked slightly cloudy.

SAQ 3.1a

3.1.2 The Ultraviolet/Visible Determinations of Glucose

Spectrophotometric methods for the determination of sugars, and particularly glucose, in the blood of diabetic patients have been much studied by clinical and biochemical analysts, while the determination of sugars is also of interest in the food industry. Although titrimetric methods were originally used, colorimetric methods were developed as early as 1920 with spectroscopy being applied to the measurement of the developed colour in the 1940s.

These early methods relied on the ability of glucose to reduce copper(II) or ferricyanide solutions, with reagents being added to the system to produce characteristically coloured complexes with the reduced copper(I) and ferricyanide products. Even now, a modern automated method of determining glucose in blood samples in based upon measurement of the loss of intensity of the yellow colour of the ferricyanide reagent by using a photoelectric colorimeter fitted with a flow-through cell.

However, these reduction methods are susceptible to errors arising from the presence of other reducing species, e.g. glutathione in blood. Enzyme reagents are much more specific and can be designed to be based on the formation of a coloured product or, as mentioned in Chapter 2, on the change in the UV spectrum of nicotinamide–adenine dinucleotide (NAD) as it is quantitatively reduced to NADH. The change in the spectrum is illustrated in Figure 3.1.

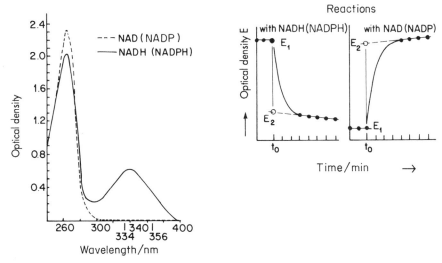

Figure 3.1   Changes in the UV absorption at 340 nm of NAD and NADH when used in enzymatic analyses

∏   An example of an enzymatic analytical procedure with the title 'UV method for the determination of sucrose and D-glucose in foodstuffs and other materials' is given below. At this stage simply make a written note of the section headings.

The method given is one detailed in a booklet of a company (Boehringer Mannheim) which supplies reagents for the enzymatic analysis of food products. Let us now check the headings you wrote down and examine the detail under each one as we go along.

(a) The *analytical principles* are outlined at the beginning of the instructions, and the *types of samples* which can be analysed are detailed in Section 10.

(b) The *amounts of samples* required are given in Section 1, with the *sample preparation methods* being detailed in Sections 10 and 11.

Sucrose/D-Glucose

Enzymatic BioAnalysis
Food Analysis

UV method
for the determination of sucrose and D-glucose in foodstuffs and other materials

Not for use in diagnostic procedures for clinical purposes
FOR *IN VITRO* USE ONLY

Cat. No. 139 041
Test-Combination for approx. 20 determinations each

For recommendations for methods and standardized procedures see references (A2, B2, C2, D2)

Principle (Ref. A 1)
The D-glucose concentration is determined before and after enzymatic hydrolysis.

Determination of D-Glucose before inversion:
At pH 7.6, the enzyme hexokinase (HK) catalyzes the phosphorylation of D-glucose by adenosine-5'-triphosphate (ATP) with the simultaneous formation of adenosine-5'-diphosphate (ADP) (1).

(1) D-Glucose + ATP $\xrightarrow{\text{HK}}$ G-6-P + ADP

In the presence of glucose-6-phosphate dehydrogenase (G6P-DH), the D-glucose-6-phosphate (G6P) formed is specifically oxidized by nicotinamide-adenine dinucleotide phosphate (NADP) to D-gluconate-6-phosphate with the formation of reduced nicotinamide-adenine dinucleotide phosphate (NADPH) (2).

(2) G-6-P + NADP $\xrightarrow{\text{G6P-DH}}$ D-gluconate-6-phosphate + NADPH + H$^+$

The NADPH formed in this reaction is stoichiometric to the amount of D-glucose and is measured by means of its light absorbance at 334, 340 or 365 nm.

Enzymatic inversion:
At pH 4.6, sucrose is hydrolyzed by the enzyme β-fructosidase (invertase) to D-glucose and D-fructose (3).

(3) Sucrose + H₂O $\xrightarrow{\text{β-fructosidase}}$ D-glucose + D-fructose

The determination of D-glucose after inversion (total D-glucose) is carried out simultaneously according to the principle outlined above.

The sucrose content is calculated from the difference of the D-glucose concentrations before and after enzymatic inversion.

The Test-Combination contains:
1. Bottle 1 with approx. 7.2 g powder mixture, consisting of: triethanolamine buffer, pH 7.6; NADP, approx. 110 mg; ATP, approx. 260 mg; magnesium sulfate, stabilizers
2. Bottle 2 with approx. 1.1 ml enzyme suspension, consisting of: hexokinase, approx. 320 U; glucose-6-phosphate dehydrogenase, approx. 160 U
3. Bottle 3 with approx. 0.5 g of lyophilisate, consisting of: citrate buffer, pH 4.6; β-fructosidase, approx. 720 U; stabilizers
4. Sucrose standard material for assay control purposes (measurement of the standard material is not necessary for calculating the results.)

Preparation of solutions
1. Dissolve contents of bottle 1 with 45 ml redist. water.
2. Use contents of bottle 2 undiluted.
3. Dissolve contents of bottle 3 with 10 ml redist. water.

Stability of reagents
The contents of bottle 1 are stable for 1 year at +4 °C.
Solution 1 is stable for 4 weeks at +4 °C, or for 2 months at –20 °C.
Bring solution 1 to 20–25 °C before use.
The contents of bottle 2 are stable for 1 year at +4 °C.
The contents of bottle 3 are stable for 1 year at +4 °C.
Solution 3 is stable for 4 weeks at +4 °C, or for 2 months at –20 °C.
Bring solution 3 to 20–25 °C before use.

Procedure
Wavelength[1]:	340 nm, Hg 365 nm or Hg 334 nm
Glass cuvette[2]:	1.00 cm light path
Temperature:	20–25 °C
Final volume:	3.020 ml

Read against air (without a cuvette in the light path) or against water
Sample solution: 4–150 μg of sucrose + D-glucose/cuvette[1]
(in 0.100–1.800 resp. 2.000 ml sample volume)

Pipette into cuvettes	Blank sucrose sample	Sucrose sample	Blank D-glucose sample	D-Glucose sample
solution 3* sample solution**	0.200 ml –	0.200 ml 0.100 ml	– –	– 0.100 ml

Mix*, incubate for 15 min at 20–25 °C or for 5 min at 37 °C, respectively, (before pipetting, warm up solution 3 to 37 °C). Add:

	Blank sucrose sample	Sucrose sample	Blank D-glucose sample	D-Glucose sample
solution 1 redist. water	1.000 ml 1.800 ml	1.000 ml 1.700 ml	1.000 ml 2.000 ml	1.000 ml 1.900 ml

Mix***, read absorbances of the solutions after approx. 3 min (A₁). Start reaction by addition of:

	Blank sucrose sample	Sucrose sample	Blank D-glucose sample	D-Glucose sample
suspension 2	0.020 ml	0.020 ml	0.020 ml	0.020 ml

Mix***, wait for completion of the reaction (approx. 10–15 min), and read absorbances of the solutions (A₂). If the reaction has not stopped after 15 min, continue to read the absorbances at 2 min intervals until the absorbance increases constantly over 2 min.

* Pipette solution 3 and sample solution each, onto the bottom of the cuvette and mix by gentle swirling. When using a plastic spatula, remove it from the cuvette only directly before measuring absorbance A₁
** Rinse the enzyme pipette or the pipette tip of the piston pipette with sample solution before dispensing the sample solution.
*** For example, with a plastic spatula or by gentle swirling after closing the cuvette with Parafilm® (registered trademark of the American Can Company, Greenwich, Ct., USA)

If the absorbance at A₂ increases constantly, extrapolate the absorbances A₂ to the time of the addition of suspension 2.

Determine the absorbance differences (A₂–A₁) for both, blank and sample. Subtract the absorbance difference of the blank from the absorbance difference of the sample.

$$\Delta A = (A_2 - A_1)_{\text{sample}} - (A_2 - A_1)_{\text{blank}}$$

The difference between $\Delta A_{\text{total D-glucose}}$ (from the sucrose sample) and $\Delta A_{\text{D-glucose}}$ (from the D-glucose sample) yields $\Delta A_{\text{sucrose}}$.

The measured absorbance differences should, as a rule, be at least 0.100 absorbance units to achieve sufficiently accurate results (see "Instructions for performance of assay").

Calculation
According to the general equation for calculating the concentrations:

$$c = \frac{V \times MW}{\varepsilon \times d \times v \times 1000} \times \Delta A \text{ [g/l]}$$

V	= final volume [ml]
v	= sample volume [ml]
MW	= molecular weight of the substance to be assayed [g/mol]
d	= light path [cm]
ε	= extinction coefficient of NADPH at
	340 nm = 6.3 [l × mmol⁻¹ × cm⁻¹]
	Hg 365 nm = 3.5 [l × mmol⁻¹ × cm⁻¹]
	Hg 334 nm = 6.18 [l × mmol⁻¹ × cm⁻¹]

---

1 The absorption maximum of NADPH is at 340 nm. On spectrophotometers, measurements are taken at the absorption maximum, if spectralline photometers equipped with a mercury vapor lamp are used, measurements are taken at a wavelength of 365 nm or 334 nm.
2 If desired, disposable cuvettes may be used instead of glass cuvettes.
3 See instructions for performance of the assay.
4 Available from Boehringer Mannheim GmbH, Biochemicals

BOEHRINGER MANNHEIM

It follows for
sucrose:

$$c = \frac{3.020 \times 342.3}{\varepsilon \times 1.00 \times 0.100 \times 1000} \times \Delta A_{\text{sucrose}} = \frac{10.34}{\varepsilon} \times \Delta A_{\text{sucrose}}$$
[g sucrose/l sample solution]

D-glucose:

$$c = \frac{3.020 \times 180.16}{\varepsilon \times 1.00 \times 0.100 \times 1000} \times \Delta A_{\text{D-glucose}} = \frac{5.441}{\varepsilon} \times \Delta A_{\text{D-glucose}}$$
[g D-glucose/l sample solution]

If the sample has been diluted during preparation, the result must be multiplied by the dilution factor F.

When analyzing solid and semi-solid samples which are weighed out for sample preparation, the result is to be calculated from the amount weighed:

$$\text{Content}_{\text{sucrose}} = \frac{c_{\text{sucrose}} \text{ [g/l sample solution]}}{\text{weight}_{\text{sample}} \text{ in g/l sample solution}} \times 100 \text{ [g/100 g]}$$

$$\text{Content}_{\text{D-glucose}} = \frac{c_{\text{D-glucose}} \text{ [g/l sample solution]}}{\text{weight}_{\text{sample}} \text{ in g/l sample solution}} \times 100 \text{ [g/100 g]}$$

1. Instructions for performance of assay

The amount of sucrose + D-glucose present in the cuvette has to be between 8 µg and 150 µg (measurement at 365 nm) or 4 µg and 80 µg (measurement at 340, 334 nm), respectively. In order to get a sufficient absorbance difference, the sample solution is diluted to yield a sugar concentration between 0.10 and 1.5 g/l or 0.05 and 0.8 g/l, respectively.

Dilution table

Estimated amount of sucrose + D-glucose per liter measurements at		Dilution with water	Dilution factor F
340 or 334 nm	365 nm		
< 0.8 g	< 1.5 g	—	1
0.8–8.0 g	1.5–15.0 g	1 + 9	10
8.0–80 g	15.0–150 g	1 + 99	100
> 80 g	> 150 g	1 + 999	1000

If the measured absorbance difference (ΔA) is too low (e.g. < 0.100), the sample solution should be prepared again (weigh out more sample or dilute less strongly) or the sample volume to be pipetted into the cuvette can be increased up to 2.000 ml (D-glucose sample) or up to 1.800 ml (sucrose sample). The volume of water added must then be reduced so as to obtain the same final volume in the cuvettes for the sample and blank. The new sample volume v must be taken into account in the calculation.

If the amount of sucrose estimated is below 0.2 g/l, the incubation time stated in the assay scheme, when sucrose is splitted by β-fructosidase, may be reduced from 15 min to 5 min.

2. Technical information

If the ratio D-glucose to sucrose in the sample is higher than e.g. 10 : 1, accuracy and precision of the sucrose determination are impaired. In this case, as much as possible of the D-glucose should be removed by means of glucose oxidase in the presence of oxygen from the air. (For details see pt. 11: Determination of sucrose and D-glucose in honey)

3. Specificity

β-Fructosidase hydrolyzes the β-fructosidic bonds in sucrose and other glycosides. If the sample contains only sucrose, it is measured specifically via D-glucose. Even in the presence of fructosanes, sucrose can be measured specifically if after enzymatic hydrolysis with β-fructosidase D-glucose and D-fructose are determined and the ratio of these monosaccharides is 1 : 1. If the D-fructose rate dominates, 2-β-fructosanes are contained in the sample.

Measurement of D-glucose and D-fructose is specific.

In the analysis of commercial sucrose, results of 100% have to be expected. In the analysis of water-free D-glucose (molecular weight 180.16) resp. D-glucose monohydrate (molecular weight 198.17), results of < 100% have to expected because the materials absorb moisture.

4. Sensitivity and detection limit

The smallest differentiating absorbance for the procedure in the determination of D-glucose is 0.005 absorbance units. This corresponds to a maximum sample volume v = 2.000 ml and measurement at 340 nm of a D-glucose concentration of 0.2 mg/l sample solution (if v = 0.100 ml, this corresponds to 4 mg/l sample solution).

The detection limit of 0.4 mg D-glucose/l is derived from the absorbance of 0.010 (as measured at 340 nm) and a maximum sample volume v = 2.000 ml.

The smallest differentiating absorbance for the procedure in the determination of sucrose (in the presence of D-glucose in the sample) is 0.010 absorbance units. This corresponds to a maximum sample volume v = 1.800 ml and measurement at 340 nm of a sucrose concentration of 1 mg/l sample solution (if v = 0.100 ml, this corresponds to 15 mg/l sample solution).

The detection limit of 2 mg sucrose/l is derived from the absorbance difference of 0.020 (as measured at 340 nm) and a maximum sample volume v = 1.800 ml.

5. Linearity

Linearity of the determination exists from 4 µg sucrose + D-glucose/assay (2 mg sucrose + D-glucose/l sample solution; sample volume v = 1.800 ml) to 150 µg sucrose + D-glucose/assay (1.5 g sucrose + D-glucose/l sample solution; sample volume v = 0.100 ml).

6. Precision

In a double determination of D-glucose using one sample solution, a difference of 0.005 to 0.010 absorbance units may occur. With a sample volume of v = 0.100 ml and measurement at 340 nm, this corresponds to a D-glucose concentration of approx. 4–8 mg/l. (If the sample is diluted during sample preparation, the result has to be multiplied by the dilution factor F. If the sample is weighed in for sample preparation, e.g. using 1 g sample/100 ml = 10 g/l, a difference of 0.04–0.08 g/100 g can be expected.)

In a double determination of sucrose using one sample solution, a difference of 0.010 to 0.015 absorbance units may occur in the presence of D-glucose in the sample. With a sample volume of v = 0.100 ml and measurement at 340 nm, this corresponds to a sucrose concentration of approx. 15–25 mg/l. (If the sample is diluted during sample preparation, the result has to be multiplied by the dilution factor F. If the sample is weighed in for sample preparation, e.g. using 1 g sample/100 ml = 10 g/l, a difference of 0.15–0.25 g/100 g can be expected.)

The following data have been published in the literature:

Fruit juice:
Sucrose: r = 1.9   + 0.033 · (c_{sucrose} in g/l)   g/l
             R = 3.3   + 0.061 · (c_{sucrose} in g/l)   g/l   (Ref. A 2.6)
D-Glucose: r = 0.42 + 0.027 · ($c_{\text{D-glucose}}$ in g/l)   g/l
             R = 1.0   + 0.042 · ($c_{\text{D-glucose}}$ in g/l)   g/l
D-Fructose: r = 0.15 + 0.033 · ($c_{\text{D-fructose}}$ in g/l)   g/l
             R = 1.05 + 0.0452 · ($c_{\text{D-fructose}}$ in g/l)   g/l   (Ref. D 2.9)

Liquid whole egg:
D-Glucose: x = 0.44 g/100 g r = 0.073 g/100 g s_r = ± 0.026 g/100 g
                              R = 0.106 g/100 g $s_{(R)}$ = ± 0.037 g/100 g
D-Fructose: x = 6.72 g/100 g r = 0.587 g/100 g s_r = ± 0.207 g/100 g
                              R = 0.748 g/100 g $s_{(R)}$ = ± 0.264 g/100 g
Sucrose: x = 43.32 g/100 g r = 1.722 g/100 g s_r = ± 1.033 g/100 g
                              R = 4.268 g/100 g $s_{(R)}$ = ± 1.501 g/100 g
For further data see references   (Ref. B 2.4)

Wine:
r = 0.056 · x_i
R = 0.12 + 0.076 x_i
x_i = D-glucose resp. D-fructose content in g/l   (Ref. D 2.17, 2.18)

7. Recognizing interference during the assay procedure

7.1 If the conversion of D-glucose has been completed according to the time given under 'determination', it can be concluded in general that no interference has occurred.

7.2 On completion of the reaction, the determination can be restarted by adding D-glucose (qualitative or quantitative): if the absorbance is altered subsequent to the addition of the standard material, this is also an indication that no interference has occurred.

The reaction cannot be restarted with sucrose as, subsequent to altering the reaction conditions from pH 4.6 to pH 7.6 ('change of the buffer'), sucrose is no longer cleaved.

7.3 Operator error or interference of the determination through the presence of substances contained in the sample can be recognized by carrying out a double determination using two different sample volumes (e.g. 0.100 ml and 0.200 ml): the measured differences in absorbance should be proportional to the sample volumes used.

When analyzing solid samples, it is recommended that different quantities (e.g. 1 g and 2 g) be weighed into 100 ml volumetric flasks. The absorbance differences measured and the weights of sample used should be proportional for identical sample volumes.

The use of 'single' and 'double' sample volumes in double determinations is the simplest method of carrying out a control assay in the determination of sucrose.

7.4 Possible interference caused by substances contained in the sample can be recognized by using an internal standard as a control: in addition to the sample, blank and standard determinations, a further determination can be carried out with sample **and** standard solution in the same assay. The recovery can then be calculated from the absorbance differences measured.

7.5 Possible losses during the determination can be recognized by carrying out recovery tests: the sample should be prepared and analyzed with and without standard material. The additive should be recovered quantitatively within the error range of the method.

8. Reagent hazard
The reagents used in the determination of sucrose and D-glucose are not hazardous materials in the sense of the Hazardous Substances Regulations, the Chemicals Law or EC Regulation 67/548/EEC and subsequent alteration, supplementation and adaptation guidelines. However, the general saftey measures that apply to all chemical substances should be adhered to.

After use, the reagents can be disposed of with laboratory waste, but local regulations must always be observed. Packaging material can be disposed of in waste destined for recycling.

9. General information on sample preparation
In carrying out the assay:
Use **clear, colorless and practically neutral liquid samples** directly, or after dilution according to the dilution table, and of a volume up to 2.000 ml (D-glucose), resp. 1.800 ml (sucrose);
Filter **turbid solutions**;
Degas **samples containing carbon dioxide** (e.g. by filtration);
Adjust **acid samples** to approx. pH 8 by adding sodium or potassium hydroxide solution;
Adjust **acid and weakly colored samples** to approx. pH 8 by adding sodium or potassium hydroxide solution and incubate for approx. 15 min;
Measure **'colored' samples** (if necessary adjusted to pH 8) against a sample blank (= buffer or redist. water + sample), adjust the photometer to 0.000 with the blank in the beam;
Treat **'strongly colored' samples** that are used undiluted or with a higher sample volume with polyvinylpolypyrrolidone (PVPP) or with polyamide, e.g. 1 g/100 ml;
Crush or homogenize **solid or semi-solid samples**, extract with water or dissolve in water and filter if necessary; resp. remove turbidities or dyestuffs by Carrez clarification;
Deproteinize **samples containing protein** with Carrez reagents;
Extract **samples containing fat** with hot water (extraction temperature should be above the melting point of the fat involved). Cool to allow the fat to separate, make up to the mark, place the volumetric flask in an ice bath for 15 min and filter; alternatively clarify with Carrez-solutions after the extraction with hot water.

Carrez clarification:
Pipette or weigh sufficient quantity of the sample into a 100 ml volumetric flask and add approx. 60 ml redist. water. Subsequently, carefully add 5 ml Carrez-I-solution (potassium hexacyanoferrate(II) (ferrocyanide), 85 mmol/l = 3.60 g $K_4[Fe(CN)_6]$ x 3 $H_2O/100$ ml) and 5 ml Carrez-II-solution (zinc sulfate, 250 mmol/l = 7.20 g $ZnSO_4$ x 7 $H_2O/$ 100 ml). Adjust to pH 7.5–8.5 with sodium hydroxide (0.1 mol/l; e.g. 10 ml). Mix rigorously after each addition. Fill the volumetric flask to the mark, mix and filter.

Samples containing protein should not be deproteinized with perchloric acid or with trichloroacetic acid in the presence of sucrose and maltose as these disaccharides are partially or fully hydrolized with the release of D-glucose. The Carrez-clarification is recommended for normal use.

10. Application examples
Determination of sucrose and D-glucose in fruit juices and similar beverages.
Filter turbid juices (alternatively clarify with Carrez reagents) and dilute sufficiently to yield a sucrose and D-glucose concentration of approx. 0.1–1.5 g/l. The diluted sample solution can be used for the assay even if it is colored. Only strongly colored juices which are used undiluted for the assay because of their low sucrose and D-glucose content should be decolorized. In that case proceed as follows:
Mix 10 ml of juice and approx. 0.1 g of polyamide powder or polyvinylpolypyrrolidone, stir for 1 min and filter. Use the clear, slightly colored solution for the assay.

Determination of sucrose and D-glucose in wine
Wine should be pretreated as described for "fruit juices". Even strongly colored sweet wines need not be decolorized. (See also pt. 11.)

Determination of sucrose and D-glucose in beer
To remove the carbonic acid, stir approx. 5–10 ml of beer in a beaker for approx. 30 s with a glass rod or filter through a folded filter. The largely CO_2-free sample can be used undiluted for the assay.

Determination of sucrose in sweetened condensed milk and ice-cream
Accurately weigh approx. 1 g sample into a 100 ml volumetric flask, add approx. 60 ml water and incubate for 15 min at approx. 70°C; shake from time to time. For clarification, add 5 ml Carrez-I-solution (3.60 g potassium hexacyanoferrate-II, $K_4[Fe(CN)_6]$ x 3 $H_2O/100$ ml), 5 ml Carrez-II-solution (zinc sulfate, $ZnSO_4$ x 7 $H_2O/100$ ml) and 10 ml NaOH (0.1 mol/l), shake vigorously after each addition, adjust to room temperature and fill up to the mark with water, filter. Use the clear, possibly slightly opalescent solution diluted according to the dilution table for the assay.

Determination of sucrose and D-glucose in jam
Homogenize approx. 10 g of jam in an electric mixer. Accurately weigh approx. 0.5 g of the homogenized jam into a 100 ml volumetric flask, mix with water, and fill up to the mark. Filter through a rapidly filtering fluted filter. Discard the first 5 ml of the filtrate. Use the clear filtrate diluted according to the dilution table, if necessary, for the assay.

Determination of sucrose in chocolate
Accurately weigh approx. 1 g of chocolate, finely grated, into a 100 ml volumetric flask, add approx. 70 ml water, and heat in a water-bath at 60–65 C for 20 min. Shake from time to time. After the chocolate has been completely suspended, allow to cool and fill up to the mark with water. To separate the fat, place in a refrigerator for at least 20 min. Filter the cold solution through a filter paper which has been moistened with the solution. Discard the first few ml of the filtrate. Use the clear filtrate diluted according to the dilution table, if necessary, for the assay.

Alternatively clarify with Carrez reagents (see pt. 9).

Determination of sucrose and D-glucose in (roasted) coffee
Accurately weigh approx. 1 g ground coffee into a 100 ml volumetric flask and add 60 ml hot water (90 C). Stir for 5 min on a magnetic stirrer. Allow to cool to room temperature and remove the magnetic rod. Clarify with Carrez reagents for separation of dyes as stated for "sweetened condensed milk and ice-cream" (see above). Use the clear, possibly slightly colored filtrate for the assay: v = 0.100 ml when analyzing raw coffee, and v = 0.500 ml when analyzing roast coffee (the altered sample volume has to be taken into account in the calculation).

11. Special sample preparation for the determination of sucrose in the presence of excess D-glucose
Determination of sucrose and D-glucose in honey
Thoroughly stir the honey with a spatula. Take approx. 10 g of the viscous (or crystalline) honey, heat in a beaker for 15 min at approx. 60 C, and stir occasionally with a spatula (there is no need to heat liquid honey). Accurately weigh approx. 1 g of the liquid sample into a 100 ml volumetric flask. Dissolve at first with only a small portion of water, and then fill up to the mark.

a) *Determination of D-glucose*
Dilute the 1% honey solution in a ratio of 1: 10 (1 + 9) and use for the assay.

b) *Determination of sucrose*
If the estimated sucrose content in the honey lies between 5 and 10%, dilute the 1% solution in a ratio of 1:3 (1 + 2) and use for the assay.

BOEHRINGER MANNHEIM

If the estimated sucrose content in the honey lies between 0.5 and 5%, the excess of D-glucose should be removed as much as possible before sucrose is determined. D-glucose is oxidized to D-gluconate in the presence of glucose oxidase (GOD) and oxygen from the air.

$$\text{D-Glucose} + H_2O + O_2 \xrightarrow{\text{GOD}} \text{D-gluconate} + H_2O.$$

Hydrogen peroxide is destroyed by catalase:

$$2\,H_2O_2 \xrightarrow{\text{catalase}} 2\,H_2O + O_2.$$

Reagents
Glucose oxidase (GOD), Cat. No. 105 139[1]
Catalase, Cat. No. 106 810[1]
Triethanolamine hydrochloride, Cat. No. 127 426[1]
$MgSO_4$ x 7 H_2O
NaOH, 4 mol/l

Preparation of solutions for 10 determinations
Enzyme solution:
Dissolve 5 mg (≙ approx. 1250 U) GOD with 0.750 ml redist. water, add 0.250 ml suspension catalase, and mix.

Buffer solution:
Dissolve 5.6 g triethanolamine hydrochloride and 0.1 g $MgSO_4$ x 7 H_2O in 80 ml redist. water, adjust to pH 7.6 with sodium hydroxide (4 mol/l), and fill up with redist. water to 100 ml.

Stability of solutions
The enzyme solution must be prepared daily.
The buffer solution is stable for 4 weeks at 4 °C.

Procedure for D-glucose oxidation

Pipette into a 10 ml volumetric flask	
buffer solution	2.000 ml
sample solution (up to approx. 0.5% D-glucose)	5.000 ml
enzyme solution	0.100 ml
Pass a current of air (O_2) through the mixture for 1 h; during the oxidation process **check the pH with indicator paper and, if necessary, neutralize the acid formed with NaOH.**	

To inactivate the enzymes GOD and catalase, place the volumetric flask in a boiling water-bath for 15 min, allow to cool, and dilute to the mark with water. Mix and filter, if necessary. Use 0.500 ml of the clear solution for the determination of sucrose. Determine the residual D-glucose in a parallel assay and subtract as usual.

(c) The *reagents* are fully specified along with the methods for *preparing solutions*. You should also note that the stabilities (shelf lives) of these solutions are also given — an important point, particularly with biochemical reagents.

(d) The *analytical procedure* includes details of:

(i)   the volumes of the reagent solutions and the final volume;

(ii) the composition of the blank;

(iii) the temperature of mixing and time before measurement;

(iv) the cell size and type, and wavelength of measurement.

∏   Can you indicate how this procedure differs in principle from that detailed for the iron determination in Section 3.1.1?

First of all, the mixing is carried out directly in the cell or cuvette. Secondly, the procedure involves the measurement of differences in absorbance. Thirdly, time is an important factor; the reaction products do not form immediately and a method of coping with a non-stable final reading is specified. Did you spot all three points?

(e) *Calculation of the analytical result* utilises previously determined values of the molar absorptivity ϵ, rather than the preparation of a set of calibration standards. This of course, limits the *accuracy of the calculated result*, since ϵ-values are subject to instrumental variations. You should also note that the procedure allows measurements to be taken at wavelengths away from λ_{max}. It is surprising that there is no indication of the standard deviation expected from the method being adopted here and there is certainly no mention of the larger variation expected when disposable cells are used!

Finally, the details given show that the method was adapted from a standard text on *Methods of Enzymatic Analysis*.

∏   Can you suggest what advantages the brochure method will have over that specified in a textbook?

Brochure methods of this type usually specify exact details which are selected to give optimum results with the reagent being employed.

Textbooks often only give general experimental details which therefore need to be adapted for the particular analysis being attempted.

SAQ 3.1b

> The success of the enzymatic determination of glucose by the method described in Section 3.1.2 is implicit in the statement that 'NADPH is stoichiometric with the amount of glucose and is determined by means of its absorption at 334, 340 or 365 nm'. Why do you think three wavelengths are specified and what are the implications in terms of the instrumental precision which is achievable?

3.1.3 Investigating a Spectroscopic Procedure

The detailed specifications of the solution conditions in a given analytical procedure are usually those which result in rapid and quantitative formation of the absorbing species, in minimising deviations from Beer's law (Section 1.5), and in the suppression of interferences from other components which are present. However, it is only by reading an account of the development of the method or by long experience that you will come to appreciate which aspects of the procedure are the critical ones in respect of achieving good reproducibility (high precision) and accuracy.

The following details of the determination of ultra-trace amounts of iron in analytical grade reagents has been abstracted from the *Analyst* (Vol. 101, pp. 974–981, 1976). This is a British journal devoted to accounts of investigations into analytical procedures, and the article referred to has been chosen for the following reasons:

(a) It illustrates how the high sensitivity of a specially developed colorimetric procedure is achieved by the use of a complex which is concentrated by solvent extraction prior to measurement of the absorption (absorbance or percentage transmittance).

(b) The investigation includes the comparison of reducing agents, the testing of optimum pH conditions, and even the influence of the time of shaking the reaction mixture.

(c) It reminds you that there are always questions to be asked about purity of reagents, and the stability of the complex which is formed.

This paper is typical of a detailed investigation of a proposed analytical procedure. It is written in the standard style used in the chemical literature, beginning with a *summary* of the work and the main achievements, which is then followed by the *introduction* which deals with the background information based on previous work in the area, and so on. In the reproduction which follows we have chosen to omit the *introduction* in order to concentrate on the analytical procedure and its assessment. For the same reason, we have omitted the *list of references*, and the *conclusions*. The latter, of course, are of great interest, but you will not miss the main conclusions if you read the *summary* at the beginning of the paper.

As you now read through the paper try and appreciate the nature and reasons for each of the investigations, and note their consequences in the standard procedure which is being adopted. For example, why is the sample and reagent mixture shaken for 30 min, and why is hydroxylammonium chloride the chosen reducing agent?

974   *Analyst, December, 1976, Vol. 101, pp. 974–981*

Spectrophotometric Determination of Trace Amounts of Iron in Pure Reagent Chemicals by Solvent Extraction as the Ternary Complex of Iron(II), 4-Chloro-2-nitrosophenol and Rhodamine B

Kyoji Tôei, Shoji Motomizu and Takashi Korenaga

Department of Chemistry, Faculty of Science, Okayama University, 3-1-1, Tsushimanaka, Okayama-shi, 700, Japan

Trace amounts of iron in pure reagent chemicals have been determined by solvent extraction - spectrophotometry with 4-chloro-2-nitrosophenol and Rhodamine B. The ternary complex of iron(II), 4-chloro-2-nitrosophenol and Rhodamine B was extracted quantitatively into toluene at about pH 4.8. The absorbance of the organic phase was measured in a glass cell of 10-mm path length at 558 nm. The apparent molar absorptivity of the ternary complex in the organic phase was 9.0×10^4 l mol^{-1} cm^{-1} at 558 nm. The ternary complex was very stable and not decomposed by addition of EDTA. By using the above procedure, trace amounts of iron (10^{-7}–$10^{-4}\%$) in alkali metal salts, alkaline earth metal salts, ammonium salts, acids and bases, etc., were determined. The standard deviations of the determinations were 1–3%.

Procedure

A 20-ml volume of the sample solution (or a smaller portion if a large amount of iron was present) was transferred by pipette into a stoppered test-tube and 1 ml of buffer solution (pH 4.8) containing 10% of hydroxylammonium chloride added, followed by 1 ml of 5×10^{-3} M Rhodamine B solution and 1 ml of 1×10^{-3} M 4-chloro-2-nitrosophenol solution in that order. The solution was well mixed and 5 ml of toluene were added. The tube containing the mixture was shaken horizontally in a shaker at a rate of 280 oscillations min^{-1} for 30 min and allowed to stand for 10 min. The absorbance of the toluene phase was measured in a glass cell of 10-mm path length at 558 nm against the reagent blank.

Results and Discussion

Effect of pH

Iron was extracted quantitatively into toluene at pH 3.9–5.3 as the ternary complex iron(II) - 4-chloro-2-nitrosophenol - Rhodamine B.   In this work, toluene was selected as the extraction solvent and the pH was adjusted to 4.8 so as to ensure greater accuracy.

Shaking Time

As shown in Fig. 1, when the reagent was added as a solution in water the ternary complex was quantitatively extracted by shaking for 20 min.   If a toluene solution containing 4-chloro-2-nitrosophenol was added, the extraction time was at least 50 min.   The use of an aqueous solution of 4-chloro-2-nitrosophenol was chosen with a shaking time of 30 min.

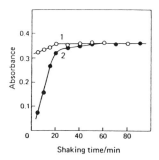

Fig. 1.   Absorbance *versus* shaking time: 1, added as an aqueous solution of 4-chloro-2-nitrosophenol; and 2, added as a toluene solution of 4-chloro-2-nitrosophenol.   Sodium chloride (manufacturer C, grade e, see Table II), 4.00 g per 20 ml.

Reducing Agent

Any iron in pure reagent chemicals was almost entirely present as the trivalent metal ion. The reduction of iron(III) to iron(II) by nine reducing agents was studied.   They were examined at concentrations of 0.5 and 5% and the results are given in Table I.   At a concentration of 0.5%, all of these reducing agents except ascorbic acid could reduce iron(III) quantitatively to iron(II).   Hydroxylammonium chloride and sulphate and sodium thiosulphate were effective at the 5% level, but 4-chloro-2-nitrosophenol was reduced and decomposed by ascorbic acid, sodium sulphite and sodium dithionite.   From these results, it was decided that hydroxylammonium salts were suitable for reducing iron(III) to iron(II) and 0.5% of hydroxylammonium chloride was used in this work.

TABLE I

SELECTION OF REDUCING AGENT

An amount of ammonium iron(III) sulphate containing 2.23 μg of iron was added to each solution.   All absorbance values were measured against a reagent blank.

Reducing agent			Absorbance	
			0.5% of reducing agent	5% of reducing agent
$NH_2OH.HCl$	..	..	0.796	0.799
$(NH_2OH)_2.H_2SO_4$	..	..	0.791	0.801
$(NH_4)_2.H_2SO_4$	..	..	0.804	0.108
Ascorbic acid	..	..	0.368	—*
Hydroquinone	..	..	0.791	0.275
Na_2SO_3	..	..	0.777	—*
$Na_2S_2O_3$	..	..	0.799	0.810
$Na_2S_2O_4$	..	..	0.783	—*
NaH_2PO_2	..	..	0.780	0.305

* 4-Chloro-2-nitrosophenol was decomposed.

Absorption Spectra

In Fig. 2, the absorption spectra of 4-chloro-2-nitrosophenol, Rhodamine B, the ternary complex iron(II) - 4-chloro-2-nitrosophenol - Rhodamine B and its reagent blank in toluene are shown. The wavelengths of maximum absorption of the ternary complex occur at 420, 558 and 700 nm. The absorption maxima at 420 and 700 nm are the absorbances of 4-chloro-2-nitrosophenol itself and the green complex of iron(II) and nitrosophenol, respectively. The absorbance of the reagent blank at 558 nm is about 0.087. The absorption spectra of sodium and potassium chloride solutions, etc., are very similar to those of distilled water. The absorbance was therefore measured at 558 nm.

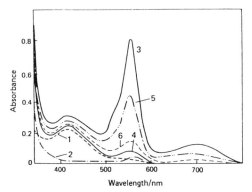

Fig. 2. Absorption spectra in toluene: 1, 2 × 10⁻⁴ M 4-chloro-2-nitrosophenol; 2, 1 × 10⁻³ M Rhodamine B; 3, ternary complex iron(II) - 4-chloro-2-nitrosophenol - Rhodamine B [8 × 10⁻⁶ M iron(II); 2 × 10⁻⁴ M 4-chloro-2-nitrosophenol; 1 × 10⁻³ M Rhodamine B]; 4, reagent blank, no iron added; 5, sodium chloride (manufacturer D, grade a, see Table II), 4.03 g per 20 ml; and 6, potassium chloride (manufacturer A, grade a, see Table II), 2.02 g per 20 ml. pH 4.8; reference, toluene.

Interference of Other Ions

It has been shown that the amounts of cobalt, nickel, copper and tin(II) ions generally present in pure reagent chemicals such as alkali metal salts, alkaline earth metal salts, acids and bases do not interfere with this method.

Monovalent anions having large molecular volumes, such as iodide, perchlorate and thiocyanide ions, could be extracted into the toluene phase as ion pairs with Rhodamine B and the absorbances of the reagent blank were greater than that of distilled water. Masking agents for iron, such as fluoride ion, interfere with the determination.

Effect of Concentration of Sample Solutions

Volumes of 100 ml of each sample solution were placed in 200-ml separating funnels and buffer solution containing hydroxylammonium chloride, Rhodamine B solution, 4-chloro-2-nitrosophenol solution and toluene were added. Iron was removed from these solutions as the ternary complex and the aqueous phase, which was free from iron, was retained and diluted as required.

A known amount of iron(II) was added to the diluted aqueous phase and the effect of the concentration of salts in the sample solutions was examined; the results are shown in Fig. 3. The results for iron in the presence of sodium and ammonium salts were not affected by the concentration of these salts. Potassium salts caused considerable interference when present at levels greater than 10% and aluminium salts at levels greater than 2%. This procedure enabled the most favourable conditions for the determination of iron in various salts to be selected.

Reagent Blank and Calibration Graph

The reagent blanks were examined by using sample solutions from which the iron had been removed. The absorption spectra of these reagent blanks were almost the same as that of distilled water to which 0.2 ml of hydrochloric acid $(1 + 1)$ per 100 ml had been added.

By adding known amounts of iron(II) ions to these iron-free sample solutions, calibration graphs were prepared and found to be straight lines with slopes almost the same as that of the calibration graph obtained by using distilled water. Accordingly, the latter graph was used to determine iron in all sample solutions.

Stability of Iron in Sample Solutions

As shown in Fig. 4, iron in 20% sodium chloride solution was not stable and the absorbance was found to decrease gradually and become zero after 5 h. When 20% sodium chloride solution to which 0.2 ml of hydrochloric acid $(1 + 1)$ per 100 ml had been added as soon as the sample solution had been prepared was used, the absorbance increased gradually, became constant after 5 h and remained at this level for at least 1 week. When the 20% sodium chloride solution to which hydrochloric acid $(1 + 1)$ had been added was heated at 80–90 °C for about 10 min constant absorbance values were obtained and they were identical with those obtained by using the solution that had been allowed to stand for 5 h.

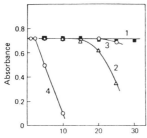

Concentration of sample solution, %

Fig. 3. Absorbance *versus* concentration of sample solutions: 1, sodium chloride; 2, potassium chloride; 3, ammonium chloride; and 4, aluminium nitrate. 8×10^{-6} M iron(II); pH 4.8; reference, reagent blank.

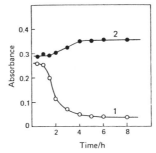

Time/h

Fig. 4. Stability of iron in sodium chloride solution: 1, no hydrochloric acid $(1 + 1)$ added; and 2, 0.2 ml of hydrochloric acid $(1 + 1)$ per 100 ml added. Sodium chloride (manufacturer C, grade e, see Table II), 4.00 g per 20 ml.

SAQ 3.1c   The molar absorptivity (ϵ_{max}) for the ternary complex of iron(II), 4-chloro-2-nitrosophenol and Rhodamine B is quoted as $9.0 \times 10^3\,m^2\,mol^{-1}$ at 558 nm. This is only a factor of four times ($4\times$) greater than that obtained with the iron(II) tripyridyltriazine (TPTZ) complex, yet the typical iron concentration levels measured are at least ten times lower (50 μg dm^{-3} against 500 μg dm^{-3}). How is this achieved?

3.2 BINARY AND MULTICOMPONENT SYSTEMS

The vast majority of the many thousands of UV/visible spectroscopic methods which have been developed are designed for the determination of a single component in a sample. However, one of the advantages of spectroscopy over visual colorimetry is the ability to handle the analysis of mixtures of absorbing species. In recent years the analysis of complex mixtures has usually involved chromatographic methods, particularly gas chromatography (GC) and HPLC, but these methods are often time consuming and require expensive reagents and solvents. With the development of computer-linked spectroscopy it is possible to undertake direct UV/visible spectroscopic analysis of mixtures by using various curve-fitting techniques. Such applications will increase in popularity as both the hardware and software required for the computer-assisted analysis are further developed.

3.2.1 Principles of the Analysis of Binary Systems

In a binary absorbing system the absorption spectra of the two components can overlap to different extents, as illustrated in Figure 3.2a.

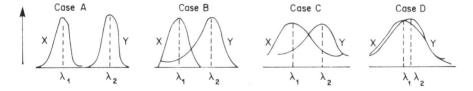

Figure 3.2a Absorption profiles for binary mixture analyses

Case A is the situation where the curves show no overlap and the two components (X and Y) are determined directly at wavelengths λ_1 and λ_2, respectively. This situation is common in the infrared region, but less common in the UV/visible region, due to the very broad absorption bands which normally occur.

Case B arises from partial overlap of the absorption spectra. While measurements at λ_2 give Y directly, any attempt to measure the absorbance of X must include some contribution due to the tail from the absorption of Y.

Case C represents the general case of overlapping absorption curves, but with the absorption maxima sufficiently displaced to allow a fairly accurate analysis by the method discussed below.

Case D is for materials with closely matching absorption curves, and is not amenable to the analysis discussed below.

For the *Case C* situation involving a mixture (M) of X at concentration c_x and Y at concentration c_y we assume that the absorbances at any wavelength are additive, so that for this mixture we can write the following:

$$A_M = A_X + A_Y \tag{3.1}$$

$$A_M = a_X c_X l + a_Y c_Y l \tag{3.2}$$

In order to use measurements of the absorbance of the mixture (A_M) to calculate the concentrations c_x and c_y in the mixture, we need to take measurements at two different wavelengths (say λ' and λ''), and we can then write:

$$A'_M = a'_X c_X l + a'_Y c_Y l \tag{3.3}$$

$$A''_M = a''_X c_X l + a''_Y c_Y l \tag{3.4}$$

These two equations can be solved for c_x and c_y provided that we have the values of the four constants (absorptivities), i.e. a'_X, a''_X, a'_Y, and a''_Y.

∏ Can you state in words what further assumptions are implied by equations (3.1) and (3.2), and hence the significance of the *a*- and *l*-values?

Equation (3.1) not only implies that the absorbances of the components in the mixture are additive, but also that they continue to obey Beer's law after being mixed together. From your knowledge of Beer's law you will know that *a* is the absorptivity for the substance(s) at the selected wavelength and *l* is the cell pathlength used for the mixture.

The solution of equations (3.3) and (3.4) is best illustrated by an example, as given below.

Example of the Analysis of a Binary Mixture

A mixture of *o*-xylene and *p*-xylene is to be analysed by UV spectroscopy in the range from 240 to 280 nm. The absorption spectra of their solutions in cyclohexane are illustrated in Figure 3.2b.

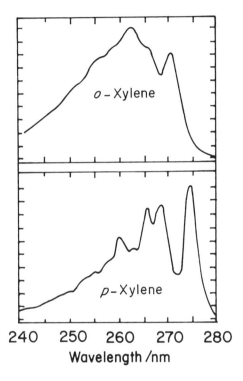

Figure 3.2b   Ultraviolet absorption spectra of xylenes in cyclohexane solution

The spectra show that *o*-xylene has absorption maxima at 263 and 271 nm, while *p*-xylene has a maximum at 275 nm. The optimum wavelengths for the analysis of the mixture are 271 and 275 nm; the 271 nm peak of *o*-xylene overlaps with a low absorption of the *p*-xylene at this wavelength, while the 275 nm peak of *p*-xylene overlaps with a low absorption due to the *o*-xylene.

Measurements on the single components and the mixture of these two wavelengths give the following:

System	Absorbance	
	271 nm (λ')	275 nm(λ'')
o-Xylene (0.40 g dm^{-3})	0.90	0.10
p-Xylene (0.17 g dm^{-3})	0.34	1.02
Mixture	0.47	0.54

The figures in the table are absorbances measured in 10 mm silica cells.

The values of the absorptivities are obtained from the single-component solutions as follows:

For *o*-xylene (component X)

$$271 \text{ nm } (\lambda') \quad a'_X = \frac{0.90}{0.40} = 2.25 \text{ dm}^3 \text{ g}^{-1} \text{ cm}^{-1}$$

$$275 \text{ nm } (\lambda'') \quad a''_X = \frac{0.10}{0.40} = 0.25 \text{ dm}^3 \text{ g}^{-1} \text{ cm}^{-1}$$

For *p*-xylene (component Y)

$$271 \text{ nm } (\lambda') \quad a'_Y = \frac{0.34}{0.17} = 2.0 \text{ dm}^3 \text{ g}^{-1} \text{ cm}^{-1}$$

$$275 \text{ nm } (\lambda'') \quad a''_Y = \frac{1.02}{0.17} = 6.0 \text{ dm}^3 \text{ g}^{-1} \text{ cm}^{-1}$$

Hence   (1) $0.47 = 2.25c_X + 2.0c_Y$

(2) $0.54 = 0.25c_X + 6.0c_Y$

This pair of simultaneous equations can be solved very simply. First eliminate c_Y by multiplying equation (1) by 3 and then subtract equation (2) from the result. This gives $c_X = 0.134$ g dm^{-3}. Calculation of c_Y then gives 0.0844 g dm^{-3}.

The final results (to two significant figures, in keeping with the experimental data) are as follows:

$$c_X = 0.13 \, \text{g dm}^{-3}$$

and

$$c_Y = 0.084 \, \text{g dm}^{-3}$$

∏ Do you fully understand why 271 nm was chosen instead of 263 nm as one of the most suitable wavelengths for measurement?

The principle reason is that at 271 nm a peak of the *o*-xylene lies in a trough of the *p*-xylene, thus avoiding measurements on the sloping side of an absorption band, which always reduces precision (Section 2.3). However, you should also note that at 275 nm the absorption coefficient of *p*-xylene ($6.0 \, \text{dm}^3 \, \text{g}^{-1} \, \text{cm}^{-1}$) is much larger than that of *o*-xylene ($0.25 \, \text{dm}^3 \, \text{g}^{-1} \, \text{cm}^{-1}$), so that this sytem approximates to Case B (Figure 3.2a), and as a result the precision of analysis, particularly for the *p*-xylene component, is quite high.

The above example is commonly used in teaching laboratories in order to demonstrate the principle of the procedure which is of general application to binary mixtures, for example with acetylsalicylic acid (aspirin) and salicylic acid.

3.2.2 Principles of the Analysis of a Multicomponent System

The above analysis can, in principle, be extended to three or more components in a mixture of absorbing species.

∏ Can you suggest what condition must hold in this case? For a three-component system, how would you select the wavelengths to make the measurements?

The conditions necessary for the above type of analysis are again as follows:

(a) the absorbances of the component in the mixture need to be additive;

(b) each component should obey Beer's law at all of the wavelengths chosen for measurement.

The number of wavelengths used for the analysis must at least equal the number of components, i.e. three wavelengths for three components, or in general, n wavelength for n components.

However, the feasibility of a particular multicomponent analysis depends on the specificity of the spectra; for a three-component system each of the components should show strong absorption in a region where the other two components show either small or zero absorption.

When these conditions hold we set up three equations and solve them in the standard manner which was previously used for the binary mixture:

$$\text{at } \lambda' \qquad A_M = a'_X c_X l + a'_Y c_Y l + a'_Z c_Z l \qquad (3.5)$$

$$\text{at } \lambda'' \qquad A''_M = a''_X c_X l + a''_Y c_Y l + a''_Z c_Z l \qquad (3.6)$$

$$\text{at } \lambda''' \qquad A'''_M = a'''_X c_X l + a'''_Y c_Y l + a'''_Z c_Z l \qquad (3.7)$$

Many three-component analyses have been reported, and also a few involving four components, but the analysis of more complex mixtures is usually not feasible in the UV/visible region. Such mixtures require either a number of analytical techniques specific to the individual component, or the mixture has to be separated by chromatographic techniques (e.g. HPLC) and then subsequently subjected to measurement by UV/visible spectroscopy.

For example, the analysis of products containing a mixture of B vitamins (say thiamin (B_1), riboflavin (B_2), pyridoxin (B_6) and nicotinamide (niacin)) usually involves the application of quantitative HPLC. However, two of the components can be separately determined by fluorescence spectroscopy (B_1 and B_2), while computer-assisted curve-fitting techniques have been developed for direct UV/visible spectroscopic analysis of the mixture. The method which has been developed depends on the change in the spectra of certain of the components with pH in order to achieve distinct absorption curves.

When a particular multicomponent analysis has been shown to be feasible and is in frequent use, it is possible to simplify the calculations involved by the use of matrix methods. However, these methods are beyond the scope of this present text and you are recommended to consult other sources if you wish to obtain details of this approach.

3.2.3 Curve-fitting Techniques

The precision of the analysis of a multicomponent system by UV/visible spectroscopy can be improved by fitting the spectrum as a whole by using least-squares criteria, rather than by simply fitting it at n different wavelengths for n components (which is the basis of the methods described in Sections 3.2.1 and 3.2.2). In practice, this involves collecting absorbance values at narrow wavelength intervals over the whole of the spectral range of interest. With modern microprocessor-controlled instruments this is readily achieved and the information is stored in digital form. However, the success of the analysis depends on the very high performance of modern spectrometers, particularly with respect to calibration checks, baseline checks, sensitivity and signal-to-noise ratio.

Again, the details are beyond the scope of this introductory course and if you are interested the relevant reference texts should be consulted.

3.3 ADDITIVE AND NON-ADDITIVE SYSTEMS; PURITY INDICES

The binary and multicomponent analyses detailed in Section 3.2 depend on a knowledge of the absorption characteristics of all of the absorbing components in the mixture. It was also assumed that the substances show additive absorbances and obey Beer's law in the mixture. In this short section, we will look first at the problem when non-additivity occurs and then at the methods of data analysis which allow us to use UV/visible spectroscopy to assess purity and to analyse the system even where there may be a background absorption of unknown origin.

3.3.1   Tests for Additivity (Binary Mixtures)

The simplest test is to measure the absorption curve of a synthetic mixture and to compare it with the predicted mixture curve by assuming additive behaviour. Figures 3.3a and 3.3b compare this approach for mixtures of two components (i.e. blue and yellow dyes) which are known to display additive and non-additive behaviours in different solvent systems. In Figure 3.3a the absorption curve falls exactly upon the points expected by adding the individual absorbances

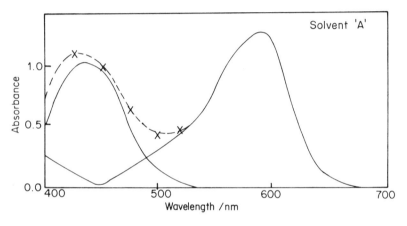

Figure 3.3a   Binary dye mixture (yellow and blue) showing additive behaviour

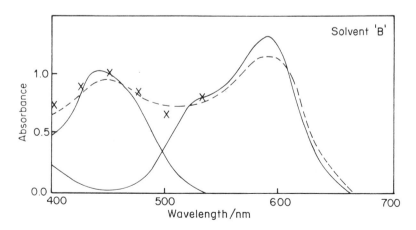

Figure 3.3b   Binary dye mixture (yellow and blue) showing non-additive behaviour

for the two dyes at any wavelength. However, in Figure 3.3b the observed absorption curve for the mixture does not coincide with the curve obtained when using the calculated values.

Non-additivity arises from chemical interaction between the components, which can be influenced by the solvent used. The problem presented in SAQ 3.3 illustrates a binary analysis in which solvent dependency of additivity is illustrated.

SAQ 3.3

A certain dye A has the same absorption spectrum in aqueous solution containing either 10% pyridine or 10% ethanol. Dye B behaves in the same way. A mixture of the two dyes, however, does not give the same absorption spectrum in the two solvents. Determine from the data below the composition of the mixture, showing how the more satisfactory data are selected.

System	Absorbance		
	450 nm	550 nm	650 nm
Dye A (1 g dm^{-3})	1.75	0.68	0.21
Dye B (1 g dm^{-3})	0.11	0.23	0.33
Mixture in ethanol–water	2.07	1.75	1.30
Mixture in pyridine–water	2.08	1.37	1.20

(Figures given in the table are absorbances measured in a 10 mm cell at the wavelengths indicated.)

To help you make a start, treat the problem as one involving a binary mixture, choose the two most appropriate wavelengths for the analysis, and then check that the result fits at the third wavelength. You will have to decide which solvent system to choose for the calculation — only one solvent gives satisfactory results.

SAQ 3.3

If a change of solvent does not eliminate non-additivity, the only recourse left to the analyst is to evaluate empirical corrections for the absorption coefficients of the components in the mixture, based on studies of synthetic mixtures covering the range of concentrations or properties likely to be met in practice.

3.3.2 Impurity and Background Absorption

Many natural products which are analysed by UV/visible spectroscopy contain impurities which contribute some background absorption or scattering over the wavelength region used in analysis. The classic example is vitamin A in fish-liver oils, where the absorption band at 328 nm is distorted due to impurity or background absorption tailing from the low-UV end of the spectrum.

A correction procedure for the analysis of this situation was developed by Drs R.A. Morton and A.L. Stubbs and is known by their names. In this technique the assumption is made that in the region of measurement the background absorption varies linearly with wavelength.

In terms of the quantities indicated on Figure 3.3c we can express the corrected absorbance as follow:

$$A_2' = A_2 - (x + y) \qquad (3.8)$$

where $(x + y)$ is the assumed background absorption. The wavelengths λ_1 and λ_3 are chosen at appropriate positions on the

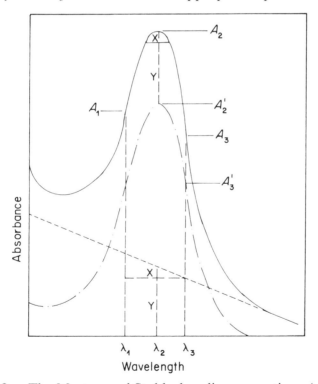

Figure 3.3c The Morton and Stubbs baseline correction: A_1, A_2 and A_3 are the absorbances at three points on the observed absorption spectrum (—); A_2' is the absorbance at λ_2 on the true absorption spectrum (–·–·–) after correcting for the background absorption (– – –) which is assumed to be linear between λ_1 and λ_3

absorption curve so that in *pure* vitamin A, $\epsilon_1 = \epsilon_3$ and $\epsilon_2/\epsilon_3 = R$ (a known ratio).

Using the property of similar triangles it is then possible to show that:

$$x = (A_1 - A_3)(\lambda_3 - \lambda_2)/(\lambda_3 - \lambda_1) \qquad (3.9)$$

and

$$y = [RA_3 - (A_2 - x)]/(R - 1) \qquad (3.10)$$

Hence the baseline is now defined.

You will have some feeling for the magnitude of the corrections by taking some typical results quoted by Morton and Stubbs for a cod-liver oil extract, as given in the following.

∏ Extraction of vitamin A from fish oil, followed by absorbance measurements of the cyclohexane extract, gave the following results:

$$\lambda_1 = 313 \, \text{nm}; \quad A_1 = 0.640$$

$$\lambda_2 = 328 \, \text{nm}; \quad A_2 = 0.712$$

$$\lambda_3 = 338.5 \, \text{nm}; \quad A_3 = 0.620$$

Given that $R = \epsilon_2/\epsilon_3 = 1.67$, show that the correction for the absorbance of the background amounts to 17.6% of the measured values at 328 nm.

Substitution in the three equations given above should give the following:

$$x = 0.008 \text{ and } y = 0.117$$

therefore $A_2' = 0.587$

So the correction at 328 nm is:

$$\frac{(0.712 - 0.587)}{0.712} = \frac{0.125}{0.712} = 17.6\%$$

The presence of background absorption in the above example can be deduced by the deviation of the A_2/A_3 or A_2/A_1 ratios from the expected value R, with such ratios being useful measures of purity, as discussed below. However, various mathematical methods have been used to handle situations where the background absorption is either flat or curved and these methods are being incorporated into curve-fitting techniques developed as part of the armoury of microprocessor-linked instrumentation.

3.3.3 Purity Checks and Purity Indices

UV/visible spectroscopy is widely used in the pharmaceutical industry as a means of quality control, mainly to check tablet dissolution rates and the concentration of UV-absorbing components. To check for impurity absorption it is necessary to have absorption data for the pure compound, and log A plots against wavelength are useful for this purpose (see Chapter 5). However, the possibility of using absorbance ratios as checks of purity has already been mentioned.

Π Suppose you were developing a method for the purity check of a synthesised vitamin A material, based on the measurements of the absorbance at λ_3 (338.5 nm) and λ_2, (328 nm). What instrumental factors would affect the measured absorbance ratio at these two wavelengths?

First of all, there would be the accuracy of *wavelength calibration*, which would need to be checked routinely.

Secondly, measurements at 338.5 nm are on the sloping side of the absorption band and are subject to variations with slit-width, which would need to be specified and kept constant.

For this second reason, purity indices are usually based on the measurement of the absorbance ratios of the peak absorbance values in the spectrum of the compound being tested, or at least the peak-to-trough absorbance values.

3.4 APPLICATIONS OF SPECTROSCOPIC ANALYSIS

∏ To remind youself of the variety of substances which can be dealt with in spectroscopic analysis, list the applications which have been mentioned in this chapter.

Iron in water

Glucose in blood

The isomer content of xylene

Vitamin A in fish-liver oil

Vitamin B mixtures

Dye mixtures in solution

Iron in reagent chemicals

∏ Can you, for each of the above, suggest the industries or institutions which would be most concerned with the above types of analysis?

The obvious answers are, in order, the water supply industry, hospitals, the petrochemical industry, the pharmaceutical industry (twice), the textile dyeing industry, and finally, the standardising laboratories of a chemical reagent supply company. However, concern for water quality arises in many different situations, while glucose content is of concern both to the individual diabetic patient and the doctor in the surgery. Thus analytical chemical techniques have to be applicable in a wide variety of environments, which are not necessarily all equipped with the best instrumentation for the analysis in question. Therefore, simple procedures are as important as those using advanced, computerised instruments. For example, although atomic absorption spectroscopy may be the preferred technique for determining iron in water, the spectroscopic method, or even a simple colorimetric analysis, based on the use of phenanthroline or TPTZ reagents, has to be available for those who do not have access to an atomic absorption spectrometer. Similarly, a combination of absorption and fluorescence spectroscopy may have to be used for vitamin B mixture analysis where an HPLC method is not available.

Summary

Many substances for determination by chemical procedures are mixtures in which the components have similar or related characteristics. UV/visible spectroscopy is capable of dealing with binary mixtures of substances with overlapping spectra by making use of Beer's law at two different wavelengths. This procedure can be extended to multicomponent systems, provided that each substance has a clearly identified maximum absorption.

Objectives

On completion of this chapter, you should now:

- be aware that there are numerous sources available of very well defined spectroscopic procedures;

- understand the need for precision and accuracy in UV/visible measurements;

- be able to follow the steps taken in a quantitative ultraviolet spectroscopic measurement;

- have a knowledge of the mathematical procedures that are used in obtaining quantitative results from a binary mixture;

- be able to carry out calculations using absorption values from binary mixtures.

4. Derivative Spectroscopy

4.1 FIRST- AND SECOND-ORDER DERIVATIVES; HIGHER-ORDER DERIVATIVES

In all of the spectra that we have seen so far we have had reasonably flat baselines and the peaks have been sufficiently well resolved for us to be able to determine λ_{max} with few problems. It is not always the case, however, that the baseline is flat and there are many occasions when you will encounter overlapping peaks. An example of such a spectrum is shown in Figure 4.1a. This is a spectrum of caffeine in

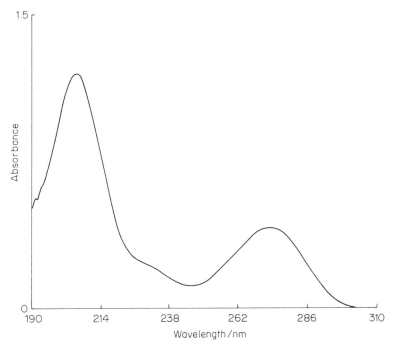

Figure 4.1a   UV spectrum of caffeine

water; it shows two prominent peaks at about 273 and 204 nm and there is also evidence of a partially resolved shoulder between these two peaks. The spectrum of caffeine in a cola drink is shown in Figure 4.1b. You will see that the peaks are much less prominent in this case and that the baseline is not flat. We would find it difficult to determine the amount of caffeine in the cola drink from this spectrum. We therefore need a method to process the spectrum. This is the basis of *derivative spectroscopy*.

The term derivative spectroscopy refers to a spectral measurement technique in which the slope of the spectrum, i.e. the rate of change of absorbance with wavelength, is measured as a function of the wavelength. Thus, the first-derivative spectrum is a plot of spectral

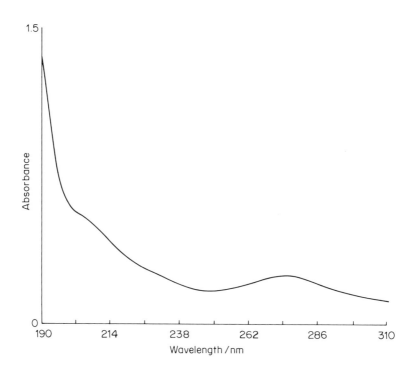

Figure 4.1b   UV spectrum of 'Pepsi Cola' showing the absorption due to caffeine

slope against wavelength and the second-derivative spectrum is itself the derivative of the first-derivative spectrum. We will illustrate this by using the zero-order curve shown in Figure 4.1c.

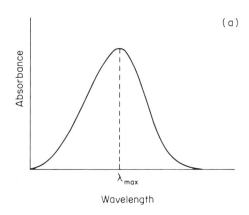

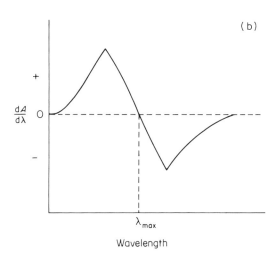

Figure 4.1c Zero-order (a) and first-order (b) derivative spectra of a Gaussian curve

We can approximate the gradient at λ_1, which we will call $dA/d\lambda_1$, by the following expression:

$$\frac{dA}{d\lambda_1} = \frac{A_2 - A_1}{\lambda_2 - \lambda_1} \tag{4.1}$$

If we carry out the calculation over many points, with $\Delta\lambda(=\lambda_2-\lambda_1)$ being small, we will obtain the smooth curve (b) which is also shown in Figure 4.1c.

I hope that you can see that $dA/d\lambda$ is zero at the wavelength corresponding to the peak maximum in the original curve.

$\prod$ Calculate the first-derivative spectrum for the spectrum of potassium permanganate shown in Figure 4.1d. Use $\Delta\lambda = 10\,nm$ and repeat the calculation every 10 nm between 420 and 630 nm.

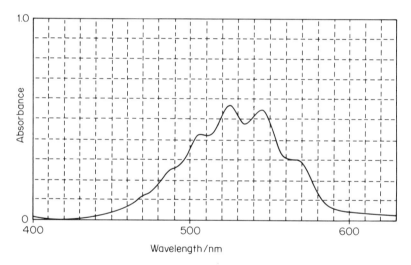

Figure 4.1d   Zero-order spectrum of potassium permanganate solution

The spectrum you should have obtained is shown in Figure 4.1e. You should compare this with the smooth first-derivative curve which was obtained by using the derivative mode of a spectrophotometer. This is shown in Figure 4.1f. You will see that in your spectrum much of the detail is missing because in sampling the spectrum only every 10 nm we miss some of the features in the original spectrum.

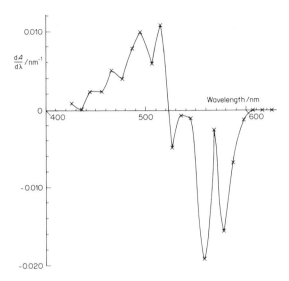

Figure 4.1e First-derivative curve, using $\Delta\lambda = 10$ nm, obtained from Figure 4.1d

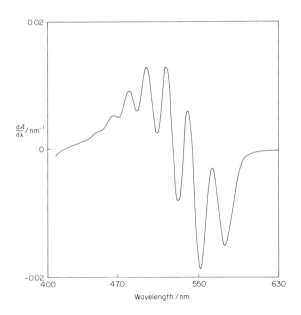

Figure 4.1f First-derivative spectrum of potassium permanganate from Figure 4.1d, obtained instrumentally

The second-derivative spectrum can be obtained by plotting the slope of the first-derivative curve, i.e. $d^2A/d\lambda^2$, against the wavelength. This is shown in Figure 4.1g for the original spectrum shown in Figure 4.1c. Higher-order derivatives are obtained in the same way and so, for example, the third-order spectrum can be found by measuring the slope of the second-order spectrum. Again, the third-order spectrum passes through zero at the wavelength corresponding to λ_{max}. This illustrates a method for the accurate measurement of λ_{max}

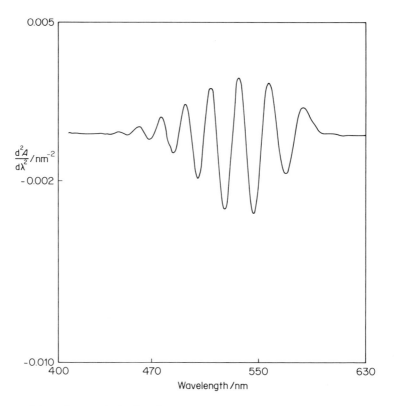

Figure 4.1g   Second-derivative spectrum of potassium permanganate obtained from Figure 4.1d

In contrast, the second- and higher even-order derivatives appear as sharpened representations of the original spectrum in which the position of the peak minimum in the second-order spectrum corresponds to the position of λ_{max} in the original spectrum. The even-order derivative peak has a sign which alternates with increasing order. It is accompanied by satellite artefact peaks, the number of which, in principle, is equal to the order of the derivative.

4.1.1 Advantages and Disadvantages of Derivative Spectra

The even-order derivative spectra have two major advantages over the zero-order spectrum, namely greater resolution and the reduction of background interferences.

Resolution

The most useful feature of derivative curves is the dramatic reduction in the bandwidth as the even-order increases. This is the basis of resolution enhancement in spectroscopic analysis. The percentage reduction in the bandwidths tends to decrease with increasing order, as can be seen in the table shown below for a Gaussian bandshape.

Derivative order	0	2	4	6	8	10
Zero-order bandwidth (%)	100	53.2	40.6	34.1	30.0	26.1

It can therefore be seen that there is little advantage to be gained for using derivatives with orders higher than four or six.

The magnitude of the resolution enhancement depends on the derivative order and the bandshape, as well as the relative bandwidths and heights of the overlapping peaks.

A disadvantage of the even-order derivatives is the generation of satellite artefacts, whose number and amplitude increases with an increase in the derivative order, as stated above. This is illustrated in Figure 4.1h. The relative magnitudes and displacements of the

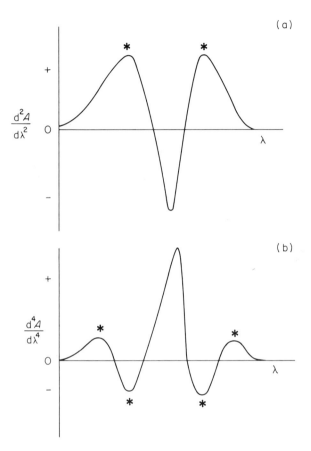

Figure 4.1h   Second- (a) and fourth- (b) derivative curves for the
Gaussian curve of Figure 4.1c; note that * indicates the satellite peaks

satellites from the central peak are given in the table below, where the
displacement from the centre is given in terms of the bandwidth, and
the relative magnitude is given as the percentage of the peak height of
the central peak.

Derivative order	First satellite		Second satellite	
	Displacement	Magnitude (%)	Displacement	Magnitude (%)
2	1.72	44.6	–	–
4	1.35	61.9	2.86	11.6

The results in the table show that these satellite peaks may interfere with adjacent peaks, and derivative spectroscopy is therefore of limited use for resolution enhancement of closely overlapping UV/visible spectral bands.

Background Interference Reduction

If the background interference due to a matrix approximates to a linear function of the wavelength, then the second-order derivative spectrum will remove it from the spectrum. This is illustrated in Figure 4.1i, where the absorption band is represented as $G(\lambda)$ and the baseline interference is given by $P(\lambda)$.

In cases where the interference can be described by a third-order polynomial the fourth derivative would satisfactorily remove its contribution.

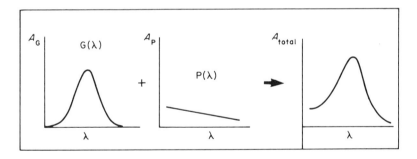

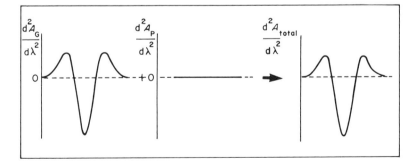

Figure 4.1i   Effect of the second derivative in discriminating against a linear matrix interference $P(\lambda)$ superimposed on a Gaussian band $G(\lambda)$

Signal-to-Noise Ratio

The other major problem in derivative spectroscopy is that the signal-to-noise ratio (S/N) degrades progressively with increasing order of the derivative. This occurs because the noise, even when small, contains by far the sharpest features in the spectrum. The S/N in higher-order derivative spectra can be improved by careful optimisation of the scan speed, the spectral slit width and the response time. The decrease in signal-to-noise ratio can further be minimised by averaging the derivative calculation over a wavelength region ($\Delta\lambda$) which is large compared with the noise structure. In unfavourable cases the S/N may decrease by a factor of about two in each successive derivative order.

Use of large values of $\Delta\lambda$ can cause distortion in the shape of the derivative spectrum, however, and so when dealing with real spectra we need to make a compromise which minimises the distortion to the spectrum but which also gives a satisfactory S/N. The optimum value of $\Delta\lambda$ and the order of the derivative will vary with the application and must be determined by trial and error. In practice, $\Delta\lambda$ should be set as high as the required resolution permits in order to obtain a good S/N.

When resolving a single band from a broad, structureless background it is best to use a relatively large value of $\Delta\lambda$ and a high derivative order so as to achieve a good S/N. When resolving several bands, however, $\Delta\lambda$ must be set somewhat smaller in order to avoid potential additional distortion.

4.1.2 Peak-height Measurement

The methods for measuring the peak height can be illustrated by reference to a spectrum of an absorption band superimposed on a sloping baseline. This is the sort of spectrum which is obtained from turbid solutions. The interference due to the baseline is progressively reduced by using higher derivatives.

The derivative peak amplitude (D) is usually measured in mm to either the long-wavelength-peak satellite (D_L) or the short-wavelength-peak satellite (D_S), as illustrated in Figure 4.1j. Alternatively, the peak-tangent baseline, D_B, or the peak-derivative zero, D_Z, could be used. Whichever method is used, the total

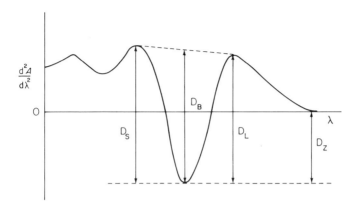

Figure 4.1j Possible ways to measure the intensity of a derivative peak; the meanings of D_S, D_L, D_B, and D_Z are given in the text

derivative amplitude (D_T) is the algebraic sum of the contributions from all of the components contributing to the peak.

For example, in a two-component system of an analyte X and a matrix M we know from Beer's law that at a wavelength λ the total absorbance, A_T, is the sum of the absorbances due to the analyte, A_X, and the matrix, A_M, as follows:

$$A_T = A_X + A_M \tag{4.2}$$

and therefore for any given second-derivative measurement we can write:

$$\frac{d^2 A_T}{d\lambda^2} = \frac{d^2 A_X}{d\lambda^2} + \frac{d^2 A_M}{d\lambda^2}$$

or
$$D_T = D_X + D_M \tag{4.3}$$

The matrix contribution, D_M, is minimised by choosing the appropriate derivative measure. This can be obtained from interaction graphs of all derivative amplitudes for D_X, at a constant level of X, against matrix concentration, which should vary from 0 to 150% of its concentration in the sample. The derivative measure selected is one where the graph is most closely parallel to the concentration axis. Deviations from this parallel line indicate an interaction and if the

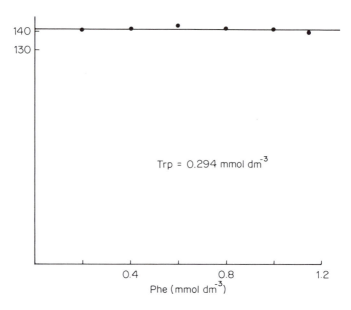

Figure 4.1k   Interaction plot for tryptophan (Trp) in the presence of
phenylalanine (Phe)

extent of interaction is unacceptable then a higher derivative may
reduce this sufficiently for an accurate quantitative analysis to be
achieved. A typical interaction graph is shown in Figure 4.1k. In this
example, we were trying to determine the concentration of the amino
acid tryptophan in the presence of phenylalanine. The derivative
measure, D_L, for tryptophan at 256 nm varies very little for changes in
the concentration of the phenylalanine from 0 to 1.4×10^{-3} mol dm^{-3}.

4.1.3   Quantitative Analysis Using Derivative Spectra

For quantitative analysis, if the analyte obeys Beer's law then the
amplitudes of the derivative spectra are also linearly related to the
analyte concentration.

Since $A = \epsilon cl$

then $$\frac{\mathrm{d}A}{\mathrm{d}\lambda} = \frac{\mathrm{d}\epsilon}{\mathrm{d}\lambda} cl \quad \text{1st derivative} \qquad (4.4)$$

and $$\frac{\mathrm{d}^2 A}{\mathrm{d}\lambda^2} = \frac{\mathrm{d}^2 \epsilon}{\mathrm{d}\lambda^2} cl \qquad \text{2nd derivative} \qquad (4.5)$$

SAQ 4.1

Standard solutions of potassium permanganate were prepared with the following concentrations:

Solution	A	B	C	D
Concentration (mg dm^{-3})	10.0	20.0	30.0	40.0

The second-order derivative spectra of these standards are shown in Figure 4.1l.

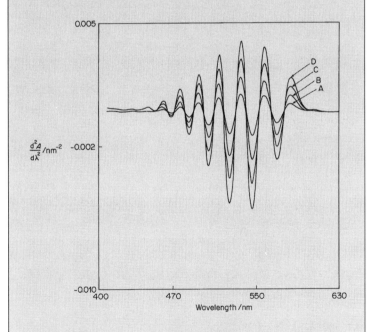

Figure 4.1l Second-order derivative spectra of potassium permanganate standard solutions →

SAQ 4.1
(Contd)

> (i) Show that these standards obey Beer's law.
>
> (ii) Use the second-order derivative spectrum of $KMnO_4$ shown in Figure 4.1g to calculate its concentration.

The factors which may affect derivative measurements are similar to those we have already met for the zero-order spectra. In particular, the following are significant:

(a) spectral band width;

(b) the sample concentration, which will determine the absorbance level and so produce high or low signal-to-noise ratios;

(c) the wavelength region in which the sample absorbs, again affecting the signal-to-noise ratio;

(d) the scanning speed, which will determine the data sampling interval ($\Delta\lambda$) for derivative measurements.

The effect of scan speed on $\Delta\lambda$ for the Varian DMS 200 spectrophotometer is illustrated in Table 4.1. The data in this table show that it is futile to attempt to obtain derivative spectra by using very fast scan speeds of 500 or 1000 nm min^{-1} over a wide spectral range.

Table 4.1 The effects of scan speed on the data sampling interval $\Delta\lambda$

Scan speed (nm min^{-1})	$\Delta\lambda$ Interval (nm)
20	0.2
50	0.5
100	1.0
200	2.0
500	5.0
1000	10.0

4.2 INSTRUMENTATION FOR DERIVATIVE SPECTROSCOPY

A variety of methods is available for obtaining derivative spectra. The two most common methods are wavelength modulation and numerical differentiation.

4.2.1 Wavelength Modulation

Several procedures have been developed for wavelength modulation. In some of these, a wavelength interval, corresponding to $\Delta\lambda$, of a few nm, is swept repeatedly as the spectrum is scanned in the normal way. The amplitude of the resulting signal from the detector is then a good approximation to the wavelength derivative. The repetitive scan can be obtained by a number of mechanical methods, including vibration or oscillation of a mirror, the slit or the monochromator.

A second method for achieving wavelength modulation involves the use of two dispersing systems (monochromators) arranged in such a way that two beams of slightly different wavelengths (usually 1 or 2 nm) fall on the sample and its detector alternatively. In this type of instrument no reference beam is used. If the two beams are scanned simultaneously the ordinate (y-axis) will display the difference ΔA between the absorbances of the two wavelengths and the x-axis will represent the mean wavelength. Since $\Delta\lambda$ is constant, the spectrum that is displayed will effectively be the first derivative. A schematic diagram of a dual-wavelength instrument is shown in Figure 4.2.

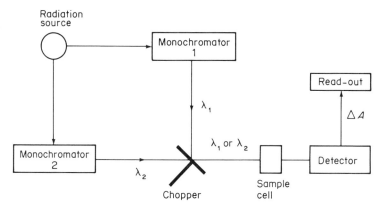

Figure 4.2 Schematic diagram of a dual-wavelength spectrophotometer; the chopper causes λ_1 and λ_2 to pass through the sample cell alternately

∏ Can you think of any other possible applications of a dual-wavelength instrument, other than for obtaining derivative spectra?

I could think of two possible uses. It could be used without scanning to measure two components, recording their variation with time in, for example, a kinetics experiment or for process control. One wavelength could be fixed at a wavelength where negligible absorption occurs, while the other is scanned, thus allowing corrections for variations in the source intensity comparable to a double-beam instrument, but by only using a single cuvette. It could also be used to study turbid solutions where the apparent pathlength is uncertain because of the effects of the suspended material. If the instrument is again set up with one wavelength fixed, while the other is scanned, the effects of changes in the effective pathlength will be eliminated.

4.2.2 Numerical Differentiation

In order to carry out numerical differentiation it is necessary to have a spectrophotometer which is microprocessor controlled. A normal, zero-order spectrum of the sample is scanned and the data are then stored digitally. The microprocessor then carries out the same calculation as you did in Section 4.1. It takes adjacent data points, separated by a wavelength, $\Delta\lambda$, and calculates the difference in the two absorbance values, ΔA. It is then easy to calculate $\Delta A/\Delta\lambda$, which is the first derivative of the original spectrum. In the example that you tried you used $\Delta\lambda = 10\,nm$ and carried out the calculation every 10 nm, and you should have seen that your result was a reasonable approximation to that obtained by using the spectrophotometer, in which data points were calculated every nm.

In Section 4.1.3, and in Table 4.1, we saw the effects of scan speed on the value of $\Delta\lambda$, namely the faster the scan speed then the greater the value of $\Delta\lambda$. How can this be explained? In most microprocessor-controlled instruments, we do not calculate $dA/d\lambda$ or $d^2A/d\lambda^2$. In fact, the derivative that is obtained is dA/dt or d^2A/dt^2, i.e., the spectrophotometer actually differentiates the changing absorbance signal against time (t), rather than wavelength. If the spectrophotometer has a constant scanning speed the two functions are equivalent, but choice of the scanning speed will directly effect the intensity of the derivative spectra. A faster scanning speed will give a more rapidly changing absorbance signal and produces an enhanced effect when the derivative is calculated. For a given response, however, the resolution of the spectrum deteriorates as the scan speed is increased and so information will be lost at high scanning rates. The most useful derivative spectra are recorded at relatively high scan speeds, i.e. usually more than about $60\,nm\,min^{-1}$.

It is important, therefore, that the same scanning speed is used for all of the standards and any unknown sample when using derivative spectroscopy for quantitative analysis.

4.3   APPLICATIONS

The concept of derivatising spectral data has been known for over 40 years, but the derivative technique did not receive much attention until about 20 years ago. This was mainly due to the complexity of generating derivative spectral data from the early UV/visible instruments, which was carried out by the use of wavelength modulation.

With the rapid developments in microcomputer technology there has been a parallel growth of interest in the use of UV/visible derivative spectrophotometry as an analytical tool for:

(a) the enhancement of the resolution of overlapping peaks;

(b) the elimination or reduction of background or matrix absorption.

Today, first- and second-order derivatives are a standard feature of most microprocessor-controlled UV/visible spectrometers, with recent areas of application of this technique including the following:

Clinical chemistry

Pharmaceutical science

Biochemistry

Inorganic and organic chemistry

Within these broad areas there is a wide diversity of applications. We have already mentioned minimisation or elimination of background absorption but derivative spectroscopy can also be used for:

(a) the characterisation of individual pure components;

(b) the study of homologous and isomeric series of compounds;

(c) the quantitative determination of trace components;

(d) the characterisation of commercial materials and natural substances.

It is, however, the Life Sciences area that has seen the most increasing use of derivative UV/visible spectrophotometry. Analyses in the clinical, pharmaceutical and biochemical areas generally have to be carried out in adverse conditions, i.e. in the presence of strongly interfering background matrices. The list below is just a sample of typical analyses in the Life Sciences field which can be carried out by using derivative methods:

(a) the analysis of colouring agents in pharmaceuticals;

(b) determination of heroin and morphine;

(c) stability of oral vitamin K;

(d) determination of carbonylhaemoglobin in blood;

(e) determination of cephalosporins.

Summary

Derivative UV/visible spectrophotometry can lead to a more accurate determination of the wavelengths of broad-peak maxima, and of peaks which appear only as shoulders, as well as the isolation of small peaks from an interfering background. Derivative spectroscopy is quantitative since the linear relationship between absorption and concentration as defined by Beer's law is unaffected by differentiation. Unfortunately, however, the advantages of derivative spectroscopy are partly offset by the decrease in signal-to-noise ratio which may occur in some cases.

Objectives

On completion of this chapter, you should now:

● understand how derivative spectra can be obtained from a normal absorption spectrum;

● be able to carry out calculations based on the features of derivative spectra;

● appreciate the scope of derivative UV/visible spectrophotometry in a wide range of applications.

5. Qualitative Analysis and Structural Relationships

5.1 IDENTIFICATION BY ABSORBANCE PLOTS

In Chapter 1 of this present Unit we reviewed the use of colour tests in the identification of chemical species, with mention being made of the characteristic purple colour of aqueous potassium permanganate solutions, which is familiar to most students of chemistry. However, if we relied only on colour as perceived by the eye we could on occasions be misled. Any particular solution colour can actually arise from a wide range of different compounds, e.g. think of all of the solutions which might be described as yellow.

Figure 5.1 shows the visible absorption spectrum of a solution of potassium permanganate, along with the absorption curve of a solution of a soluble azo dye which appears to the eye to have the same purple colour as the permanganate solution. In viewing solutions of these two substances the eye registers identical colours because both species have an absorption band which is centred on the green region near 530 nm; the absorptions are also of approximately the same intensity.

Although the colours appear identical the shapes of the absorption curves, as recorded by the spectrometer, are obviously different, and in using absorption curves in the UV and visible regions for purposes of identification or qualitative analysis it is important not only to specify the position and intensity of the absorption bands but to also indicate in some way the shape of the absorption curve.

As we have seen previously in this Unit the position of an absorption band in the UV/visible region is usually defined by quoting the wavelength (λ) in nm, although occasionally wavenumbers ($\bar{\nu}$) in cm^{-1}

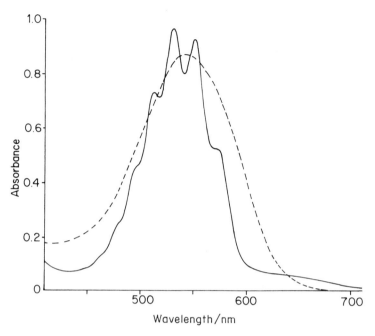

Figure 5.1   Absorption spectra of two substances with identical colours

or even frequencies (ν) in hertz are used. The relationship between these has already been given, and you may recall that it is as follows:

$$\lambda\nu = c; \quad \nu = c/\lambda; \quad \bar{\nu} = 1/\lambda$$

with the velocity of light (c) being taken as $3 \times 10^8\,\text{m s}^{-1}$.

You will also know that the intensity of any particular absorption band in the UV/visible region is usually quoted in terms of the molar absorptivity ϵ, as defined by Beer's law:

$$A = \epsilon c l$$

Here, the concentration term (c) needs to be expressed in mole terms (usually mol dm^{-3}) and the cell pathlength in mm or cm. Where the molecular mass of a substance is unknown, or the purity of the product is unspecified (e.g. with dyes and many natural products), it is common practice to use the concentration (c') expressed in g per 100 ml (%(wt/vol)) when the absorptivity is represented by the symbol $E_{1\%}^{1\ cm}$, and the Beer's law then becomes:

$$A = E_{1\%}^{1\ cm}\ c'l \text{ (with } l \text{ in cm)}$$

∏ Use the information in Figure 5.1 to evaluate the intensity of the permanganate and azo dye absorption in the following units:

Position unit	Intensity unit
λ_{max} (nm)	$E_{1\%}^{1\ cm}$
ν_{max} (cm^{-1})	ϵ_{max} $(dm^3\ mol^{-1}\ cm^{-1})$
ν_{max} Hz	ϵ_{max} $(m^2\ mol^{-1})$

You should have obtained the following values, although the actual results will depend upon the figures obtained for the absorbances at λ_{max}:

Parameter	$KMnO_4$	Azo dye
λ_{max} (nm)	533	542
$\bar{\nu}_{max}$ (cm^{-1})	18 760	18 450
ν_{max} (Hz)	5.62×10^{14}	5.53×10^{14}
$E_{1\%}^{1\ cm}$	640	435
ϵ_{max} $(dm^3\ mol^{-1}\ cm^{-1})$	1.01×10^4	1.88×10^4
ϵ_{max} $(m^2\ mol^{-1})$	1.01×10^3	1.88×10^3

The azo dye has an unstructured absorption curve with a single peak, whereas the permanganate absorption shows some structure. This may be indicated by quoting the absorption characteristics of $KMnO_4$ in the following form:

$$\lambda(nm)\ (496),\ 514,\ 533,\ 553,\ (570)$$

which indicates a triple peak with shoulders in the absorption at 496 and 570 nm.

The wavenumber values in cm^{-1} are obtained by using $\bar{\nu} = 10^7/\lambda$, with λ in nm, while the frequency values in Hz are similarly obtained from $\nu = 3 \times 10^{17}/\lambda$, again with λ in nm. It should be noted that the final unit for ϵ_{max} is in SI units derived by using c in $mol\,m^{-3}$ and l in m.

Therefore, the two purple compounds in Figure 5.1 are readily distinguished because of the characteristic structured appearance of the $KMnO_4$ spectrum. The shoulders and side bands arise as a result of the spectrometer separating the vibrational structure of the electronic transition in the MnO_4^- species. Such structure is relatively unusual in spectra of solutions in the UV/visible region, although, of course, the detailed vibrational pattern in an infrared spectrum leads to the description of such spectra as *chemical fingerprints*.

Another distinguishing characteristic is the width of the absorption band, but this is rarely quoted in tables of absorption characteristics. The $KMnO_4$ spectrum has a narrower or sharper absorption band than the dye.

∏ Can you suggest how bandwidths might be measured?

From what has been covered in earlier sections you should be aware that it is by quoting the bandwidth in nm at the half-peak height.

It should be apparent to you from this introductory discussion that in order to use UV/visible absorption data for qualitative analysis we require more than just the peak position and intensity data. Various means have been devised for distinguishing materials which have similar absorption characteristics, varying from curve shape indices to the use of logarithm of absorbance ($\log A$) plots. The changes in these quantities or plots when the solution conditions are altered, or when the chemical species undergoes reaction with an identifying reagent are also useful. Note that the absorbance ratio method of checking purity mentioned in Chapter 3 uses a simple curve-shape parameter.

As you might expect there is a relationship between absorption characteristics and chemical structure. The final part of this chapter will examine how changes in the UV/visible absorption are associated with structure in organic compounds and how these changes can be used to investigate solution reactions and equilibria.

5.2 PRESENTATION OF ABSORPTION DATA IN THE ULTRAVIOLET AND VISIBLE REGIONS

As indicated in the above introduction, the simplest method of characterising a material which has a single absorption band in the UV or visible regions is to quote the position and intensity of the absorption in suitable units. Most of the standard texts on the UV/visible spectra of organic compounds utilise tables of λ_{max} (in nm) and ϵ_{max} (in $dm^3\,mol^{-1}\,cm^{-1}$). Table 5.2 illustrate a typical table of values; you should remember, however, that the values of ϵ that are given should be divided by 10 in order to give the corresponding values in units of $m^2\,mol^{-1}$.

This type of information is only a general guide to the absorption characteristics of a group of compounds since the precise values of λ_{max} and ϵ_{max} will depend on instrumental characteristics, solvent and solution conditions, and material purity.

Full absorption spectra are generally more useful, although with all of the alternative scales that are possible for position and intensity, the methods of presentation of data are still the subject of controversy.

In some libraries the spectra are plotted by using log ϵ as the vertical intensity axis and cm^{-1} (wavenumber) as the horizontal axis to indicate position (with corresponding nm wavelength units as a parallel scale). The position and intensity of all of the major peaks in the spectrum may also be presented in tabular form. Details of the

Table 5.2 Ultraviolet absorption spectra of dienes; note that a
mixture of International Union of Pure and Applied Chemistry
(IUPAC) nomenclature and trivial chemical names is used in this table

Class of compound	Compound	λ_{max} (nm)	ϵ_{max} (dm^3 mol^{-1} cm^{-1})
Linear dienes	1,3-Butadiene	217	21 000
	Isoprene	220	23 000
	2,4-Hexadiene	227	23 000
Semicyclic dienes	β-Phellandrene	231	9 100
	Cyclohex-1-enylethylene	230	8 500
	Menthadiene	235	10 700
Cyclic dienes	Cyclopentadiene	238	3 400
	Cyclohex-1,3-diene	256	8 000
	Cyclohepta-1,3-diene	248	7 500
Polycyclic dienes	L-Pimeric acid	273	7 100
	Ergosterol	280	13 500
	7-Dehydrocholesterol	280	11 400
	7-Dehydrocholestene	280	12 700
	Cholesta-3,5-diene	235	23 000
	Cholestadienol-C	248	17 800
	Ergosterol-D	242	21 400
	Abietic acid	238	16 100

compound purity, instrument used, nature of solution, and conditions
of measurement should all be listed, since all of these are factors
which can influence the absorption curve. The use of such collections
of spectra can be helpful when attempting to apply UV/visible
spectroscopy to the analysis of compounds for which standard spectra
of your own are not available. However, when employing published
spectra you must try to avoid confusion because of the different
definitions, symbols and methods of presentation used in the various
commercial publications and literature sources. The most reliable
analyses will be the ones which utilise your own data, since you will

be aware of sample purity and solution conditions, and will be making comparisons by using spectra recorded on the same instrument (hopefully).

SAQ 5.2

Figure 5.2 is an absorption spectrum of potassium permanganate in which the four major peaks have been numbered. Values for λ_{max} and ϵ_{max} for the strongest peak are given below. Measurements were made using a 1 cm cell.

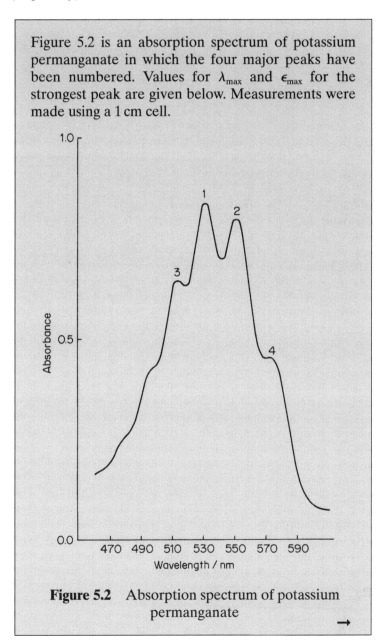

Figure 5.2 Absorption spectrum of potassium permanganate

SAQ 5.2
(Contd)

Complete the following table to show the λ_{max} and ϵ_{max} values of the three major bands.

Band	λ_{max} (nm)	ϵ_{max} (m² mol⁻¹)
1	533	1.01×10^3
2		
3		
4		

5.3   LOG A CURVES AND CURVE-SHAPE PARAMETERS

The shape of a measured absorption curve in the UV/visible region depends on the variation of absorbance with wavelength, which in turn depends on the variation of ϵ and λ. Expressed in terms of Beer's law we can write the following:

$$A_\lambda = \epsilon_\lambda c l \tag{5.1}$$

and by taking logarithms;

$$\log A_\lambda = \log \epsilon_\lambda + \log c l \tag{5.2}$$

The shape of the log A_λ curve, as distinct from the A_λ curve, is independent of pathlength and concentration. This is illustrated in Figure 5.3 in which four different solutions of potassium permanganate have been studied by using the two techniques.

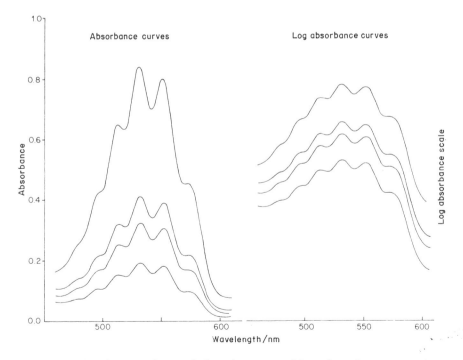

Figure 5.3   Comparison of absorbance and log absorbance spectra

If the $\log A_\lambda$ curve of a known standard compound is prepared on a transparent sheet it is possible to check for identity or purity simply by overlaying the standard curve on that of the sample being tested. A coincidence of the curve shapes is a very good test of identity. This procedure has been extensively used in the dye and pigment industry for qualitative analysis and purity checks. It is, of course, applicable to the absorption spectra of all classes of compounds, and it is recommended that where absorption spectroscopy is being used as a quality control method the absorption data of the standards should be stored in $\log A_\lambda$ form. Most modern recording UV/visible spectrometers have this facility.

Let us finally consider the relationship of the $\log A_\lambda$ method to the peak ratio method of checking purity; this latter method was mentioned in Chapter 3.

Suppose that a solution of a standard compound has two peaks with absorbance values of $A_{\lambda_1} = 0.2$ and $A_{\lambda_2} = 0.4$. For this solution, the *peak ratio* value is as follows:

$$A_{\lambda_2}/A_{\lambda_1} = 0.4/0.2 = 2:1$$

At double the concentration the peak ratio remains the same:

$$A_{\lambda_2}/A_{\lambda_1} = 0.8/0.4 = 2:1$$

With a $\log A_\lambda$ facility we would measure the difference in peak heights, as follows:

$$\log A_{\lambda_2} - \log A_{\lambda_1} = \log 2 = 0.301$$

which would be the same for both solutions.

5.4 SPECTRA–STRUCTURE CORRELATIONS IN ORGANIC CHEMISTRY

Attempts to find relationships between colour and chemical constitution have stimulated the endeavours and imagination of chemists for over 100 years. However, it was only with the introduction of quantum concepts and theoretical calculations based on wave mechanics that the complex nature of these relationships were gradually appreciated. Even today, molecular orbital calculation methods are still being developed to explain the origin of colour in chemical species such as dyes and transition metal complexes.

In this present section we shall consider some of the empirical relationships which have been useful in elucidating the structure of simple organic compounds which absorb in the accessible UV/visible (180–700 nm) regions of the spectrum.

The work of the early chemists on the colour and constitution of organic compounds introduced terms such as chromophore (used for chemical grouping which had the property of conferring colour on a substance) and auxochrome (which relates to a group not capable of producing colour on its own, but possessing the power to modify or enhance the colour produced by the chromophore). We now use the term *chromophore* to indicate a group capable of producing absorptions in the 200–800 nm range, with the *auxochrome* shifting the absorption to longer wavelengths (known as a red or bathochromic shift).

Some typical chromophoric groupings are illustrated in Table 5.4a, along with the absorption positions for simple compounds containing these chromophoric groups. This table indicates the position only of

Table 5.4a Typical chromophoric groups present in organic compounds

Group	Compound	λ (nm)
$>C=C<$	$CH_2=CH_2$	180
⬡	C_6H_6	255
$>C=O$	$(CH_3)_2CO$	277
$-N=N-$	$CH_3-N=N-CH_3$	347
$>C=S$	$(CH_3)_2C=S$	400
$-N=O$	$C_4H_9N=O$	665

the longest wavelength absorption band, but it gives no indication of the intensity of absorption (the ϵ_{max} values can vary from 10 to 10^5). For example, propanone, $(CH_3)_2CO$, shows three absorptions of differing intensity which have been assigned to three different electronic transitions.

The effect of an auxochrome (e.g. the —NH_2 group) on the characteristic absorptions can be seen in Table 5.4b.

Table 5.4b　Effect of the auxochrome —NH_2 on the λ_{max} and ϵ_{max} values of $>CH_2{=}CH_2<$ and C_6H_6

Compound	λ_{max} (nm)	ϵ_{max} ($m^2 \, mol^{-1}$)
$CH_2{=}CH_2$	180	10^3
$CH_2{=}CHNH_2$	220	10^4
(benzene ring)	255	23
(benzene ring)—NH_2	280	143

In both cases, introduction of the auxochrome leads to a higher absorption at a longer wavelength.

Combining chromophores or extending the conjugation can also significantly affect the position and intensity of the absorption, as shown in Table 5.4c.

How can we explain these effects? With conjugation, the greater the number of π electrons in the system then the less is the energy required to promote one of these electrons into the excited state, and hence the λ_{max} value increases. The delocalisation effect of conjugation reduces the gap between the ground state and the excited state. The value of ϵ_{max} tends to increase in line with the number of electrons in the conjugated system.

Table 5.4c   Effect of conjugation on λ_{max} and ϵ_{max}

Compound	λ_{max} (nm)	ϵ_{max} (m^2 mol^{-1})	λ_{max} (nm)	ϵ_{max} m^2 mol^{-1})
$CH_2{=}CH_2$	180	10^3	–	–
$CH_2{=}CH{-}CH{=}CH_2$	217	2.1×10^3	–	–
$H_2C{=}O$	180	10^3	273	1.2
$CH_2{=}CH{-}CH{=}O$	217	1.6×10^3	321	2.0

The auxochromic effect depends on the ability of the chemical group to donate electrons into the conjugated system. This has been studied most extensively with aromatic systems and the spectral shifts of monosubstituted aromatic compounds have been correlated with the electron-donating power. The electron-donating properties of some common substituents decrease in the following order:

$$O^- > NHCH_3 > NH_2, OH > Cl > CH_3 > NH_3{}^+, H$$

In this list the significant effect of protonating the NH_2 group should be noted; the proton binds the non-bonding electrons on the nitrogen of the amino group and thus prevents them from interacting with the benzene π-electron system.

SAQ 5.4

Assign from list 2 below, the appropriate values of λ_{max} and ϵ_{max} for each system given in list 1. Comment on the shifts observed relative to benzene (λ_{max} = 255 nm; ϵ_{max} = 23 m^2 mol^{-1}) in terms of the electronic properties of the auxochromes.

List 1   Absorbing systems:

A   aqueous solution of aniline;

B   solution A with 0.01 M HCl added;

C   aqueous solution of phenol;

D   solution C with 0.01 M NaOH added.

→

SAQ 5.4
(Contd)

List 2   Absorption characteristics:

System	λ_{max} (nm)	ϵ_{max} (m^2 mol^{-1})
E	255	20
F	270	140
G	280	130
H	290	230

Futher detailed discussions of the structural influences of chromophores and auxochromes on the absorption of chemical species are outside the scope of this present text. If you are interested in extending your knowledge of these effects you should study the appropriate texts listed in the Bibliography. However, we will study here the use of what are known as the Woodward rules for predicting the absorptions of diene systems as just one example of a quantitative, although empirical, correlation between UV absorption and chemical structure.

5.5 WOODWARD RULES FOR DIENE ABSORPTION

A number of important steroids contain polycyclic diene structures and Woodward and his co-workers developed certain rules for relating their UV absorptions to structural characteristics. The rules provide values for calculating wavelengths for the maximum absorptions based upon a parent diene, as a result of a red shift arising from extension of the conjugation and/or the presence of various auxochromes. Typical values for diene absorptions are given in Table 5.5. However, in order to be able to apply these rules we must be able to identify the types of structures referred to in this table.

Table 5.5 Values of λ for the various components when applying the Woodward rules for diene absorption (in EtOH solution)

Component	λ (nm)
Parent	
Acyclic diene	217
Heteroannular diene	214
Homoannular diene	253
α,β-Unsaturated carbonyl unit	222
Increment	
Double bond (extending the conjugation)	30
Alkyl group or ring residue	5
Exocyclic double bond	5
Polar group	
O (acyl)	0
O (alkyl)	6
S (alkyl)	30
Cl, Br	5
N(alkyl)$_2$	60

The total of the various components, etc. will give λ_{max}

The basic chromophore unit in the example we will consider first is 1,3-butadiene, which is regarded as the parent acyclic (or non-cyclic) diene:

$$H_2C \overset{H}{\underset{}{=}}C \text{---} C \overset{=CH_2}{\underset{H}{}} \quad \text{or} \quad$$

If the diene has saturated alkyl groups attached to it, then an additional contribution for each group is added as follows:

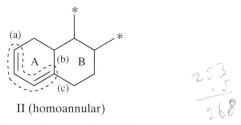

I

This is analysed in terms of the rules as follows:

Base value for I (acyclic structure) = 217 nm

For three alkyl or methyl groups (a, b, c)

add 3 × 5 = +15 nm

To give a predicted λ_{max} = 232 nm

Compared with the observed λ_{max} = 231 nm

If the diene system is contained in a single ring it is termed *homoannular*, whereas if it is spread over two rings it is said to be *heteroannular*.

II (homoannular)

For compound II we have a base value = 253 nm

The ring residues (a, b, c) are all attached to the diene system

so add 3 × 5 = +15 nm

The lower double bond in A is attached to, but outside (or exocyclic to) ring B = +15 nm

To give a predicted λ_{max} = 273 nm

Compared with the observed λ_{max} = 275 nm

Note that the groups marked with * do not contribute as they are not directly attached to the diene system.

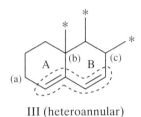

III (heteroannular)

For compound III we have a base value = 214 nm

For the ring residues (a, b, c), add 3 × 5 = +15 nm

The double bond in A is *exocyclic* to ring B = +15 nm

To give a predicted λ_{max} = 234 nm

Compared with the observed λ_{max} = 235 nm

In addition, if a ring structure such as the following occurs:

IV

then both of the double bonds are in exocyclic positions, and therefore compound IV would have a 10 nm contribution from this element of the structure.

∏   Identify the diene structure and the contributing elements in compound IV and use the Woodward rules to predict λ_{max} in the UV absorption spectrum.

Your calculations should give the following:

$$\lambda_{max} = 214 + 25 + 10 = 249\,nm$$

Note that one of the methyl groups attached to ring B influences the diene absorption, along with the four ring residues attached to the chromophore, giving in total $5 \times 5 = 25\,nm$.

Similar rules also apply for α,β-unsaturated carbonyl compounds. The complete set of tables required for using the Woodward rules also includes corrections for solvent shifts, since the electronic transition in the carbonyl group is sensitive to the polarity of the solvent.

SAQ 5.5

A compound with the following structure:

is classed as (i) an alkene if X is CH_2, and (ii) an unsaturated ketone if X is O.

For the purpose of applying the Woodward rules to predict the λ_{max} of the UV spectrum in alcohol solution we need to be able to characterise certain aspects of the structural contributions.

(A) In this context, indicate which of the following statements are *true* and which are *false*.

(i) When X is CH_2 we have a triene and the diene rules do not apply.

(ii) The compound has only *two* exocyclic double bonds.

SAQ 5.5
(Contd)

(iii) The compound has two methyl groups, both of which contribute +5 nm to the absorption.

(iv) When X is an oxygen atom (O) the double bonds of the unsaturation are in the α, β and γ, δ positions.

(B) By using the appropriate rules, show that the predicted λ_{max} values in EtOH are as follows:

(i) 269 nm for the alkene structure;

(ii) 277 nm for the ketone structure.

You may wish to try your hand at applying the Woodward rules yourself. The following structures are given with both the observed λ_{max} values and the corresponding calculated values (which you should be able to obtain without too much difficulty) in parentheses.

238 (239)   235 (234)   273 (278)

238 (239)   276 (274)   257 (256)

$(CH_3)_2C{=}CHCOCH_3$
237 (239)   232 (237)   243 (249)

254 (254)   225 (227)   241 (242)

290 (286)   240 (239)   270 (274)

294 (299)   313 (315)   278 (289)

5.6 FOLLOWING REACTIONS AND EQUILIBRIA

Although the use of UV/visible spectroscopy as a means of investigating molecular structure is limited in comparison with infrared and nuclear magnetic resonance spectroscopy, it remains an important tool for the investigation of reactions and equilibria in solution. Any change which affects the electronic properties of the absorbing species in the solution can provide a useful means of investigating the change. Examples of reactions which have been studied by UV/visible spectroscopy include the following:

Oxidation of unsaturated oils;

Methylation of amino groups in dye synthesis;

Ligand replacement reactions in complex ions;

Cis–trans isomerisation in conjugated aromatic systems;

Fast reactions studied by flash photolysis techniques.

Examples from the first and the last of these categories are given below as illustrations of the use of UV/visible spectroscopy to study reaction mechanisms and reaction kinetics, respectively.

The study of chemical equilibria is also an important application of UV/visible spectroscopy. The types of equilibria which have been studied by this method include the following:

Acid–base equilibria of indicators;

Chlorophyll–protein complexes;

Dye–surfactant complexes;

Dimerisation or aggregation behaviour;

Charge-transfer complexes.

However, certain difficult problems can occur in such equilibria studies, namely:

(i) the identification of those situations in which a stoichiometric equilibrium exists between two or more species;

(ii) the identification of the species in equilibrium;

(iii) the ability to devise methods for obtaining the absorption characteristics of the individual species.

In this present section we will restrict ourselves to these problems, although there are many other aspects which are outside the scope of this text.

5.6.1 Reaction Studies Involving a Chemical Change

It has been known for a long time that the thermal oxidation of unsaturated oils involves a free-radical mechanism in which peroxides and more complex oxidation products are formed at the double bonds in the oil molecules. The possibility of photosensitised initiation by singlet oxygen has been studied and the extent of monoperoxide formation followed by UV spectroscopy. A typical set of results which have been obtained from the photooxidation of methyl linoleate is presented in Figure 5.6.

It is now known that such a photoxidation process produces four possible monohydroperoxides, which are shown in the following reaction scheme:

∏ Using your knowledge of the criterion for UV absorption, indicate which of the products (A, B, C, or D) gives rise to the absorption at 233 nm. If the mean molar absorptivity of the absorbing species is $2.6 \times 10^3 \, \text{m}^2 \, \text{mol}^{-1}$, calculate the fraction of the original linoleate which has been converted to monohydroperoxides after an irradiation time of 90 min.

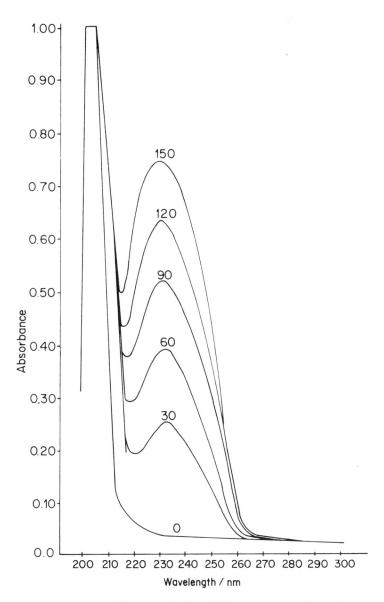

Figure 5.6   Changes in the UV spectrum of methyl linoleate/erythrosin (0.04 mol dm^{-3} ester) with irradiation time (min); solutions are diluted by 1/100 for ease of measurement

Products B and C are conjugated dienes, whereas products A and D are unconjugated species. Using the diene rules we expect B and C to have absorptions at 217 nm + 15 nm = 232 nm, i.e. close to the observed position. To measure the extent of conversion we note that the reaction mixture has been diluted by a factor of 100 and that the absorbance change after 90 min is ~ 0.5 in a 1 cm cell. This corresponds to a concentration of $0.5/2.6 \times 10^3$, or before dilution, of 100 times this value. This is equivalent to the formation of ~ 0.002 mol dm^{-3} of conjugated species. If we assume that the conjugated and unconjugated species are present in equal amounts, then the total change is ~ 0.004 mol dm^{-3} product, or $\sim 10\%$ of the original ester.

The above example illustrates the way in which UV spectroscopy can be used as an aid to following a reaction both mechanistically and quantitatively.

5.6.2 Following Fast Reactions by Kinetic Spectroscopy

A number of fast-reaction techniques utilise the changes in UV/visible absorption as a means of following the formation and/or reaction of short-lived or transient species. Some common examples of these are stopped-flow and pulsed irradiation techniques.

In these systems, lasers are employed for very short flash irradiations, with the formation and decay of the various species being measured at fixed UV wavelengths. From the decay curves that are obtained the kinetics of the process can be postulated. However, the procedures and details required for these studies is outside the scope of this present text. However, by now you should be well aware that UV/visible spectroscopy is a very valuable tool for the analyst in all forms of chemical and biochemical studies for both qualitative and quantitative determinations.

Summary

The wavelengths and intensities of the ultraviolet/visible radiation absorbed by a substance is related to its chemical structure, with the spectra of organic compounds being particularly sensitive to the presence of unsaturation and groups with lone-pair electrons. By assessing the contributions due to the various structural features and functional groups of an organic molecule it is possible to calculate (with a high degree of success) the wavelengths at which many substances exhibit their peak absorptions.

Objectives

On completion of this chapter, you should now:

● be aware that different chemical compounds can have an apparent identical colour but different absorption spectra;

● have learnt that well defined chemical features have a major influence on the nature of the UV/visible spectrum;

● be able to carry out simple summation procedures for calculating the wavelength of maximum absorption for various compounds;

● understand the importance of spectral measurements in studying chemical reactions;

● have a knowledge of the value of UV/visible spectroscopy in qualitative and quantitative analysis.

Self-assessment Questions and Responses

SAQ 1.1a

> Wavenumber ($\bar{\nu}$) values in cm^{-1} units are calculated by taking the reciprocal of the wavelength (λ) values, and multiplying these by an appropriate factor to allow for the conversion of units.
>
> (i) What is the relationship between wavelength values in nm and wavenumbers in cm^{-1} units?
>
> (ii) Similarly, what is the relationship between wavenumber values in cm^{-1} units and frequency values in hertz (or s^{-1})?
>
> Use these two relationships to calculate the wavenumber and the frequency of yellow radiation of wavelength 575 nm. Check your values against the wavenumber and frequency scales of Figure 1.1a.

Response

(i) Wavenumber = 1/wavelength or $\bar{\nu} = 1/\lambda$

Allowing for units:

wavenumber (cm^{-1}) = 10^7/wavelength (nm)

(ii) $\bar{\nu} = 1/\lambda = \nu/c$

wavenumber = frequency/velocity (c)

wavenumber (cm^{-1}) = frequency (hertz) $\times$ 100/velocity (ms^{-1}) or frequency = wavenumber $\times$ velocity/100

Applying these relationships to yellow radiation of wavelength 575 nm

$$\bar{\nu} = 10^7/575 = 17\,391\,cm^{-1} = 17\,391 \times 10^2\,m^{-1}$$

$$\nu = 17\,391 \times 10^2 \times 3 \times 10^8 = 5.2 \times 10^{14}\,s^{-1}\,(hertz)$$

These values agree with the scales in Figure 1.1a.

SAQ 1.1b	Using your knowledge of the colours of the common reagent solutions listed below, identify the solutions corresponding to the spectra A to E in Figure 1.1d.

Solution	Reagent (concentration)
1	aqueous copper sulfate solution $(0.4\,mol\,dm^{-3})$;
2	aqueous copper sulfate solution $(0.04\,mol\,dm^{-3})$;
3	aqueous potassium dichromate solution $(0.02\,mol\,dm^{-3})$;
4	potassium dichromate $(100\,mg\,dm^{-3})$ in dilute sulfuric acid $(0.005\,mol\,dm^{-3})$;
5	aqueous potassium permanganate solution $(5 \times 10^{-4}\,mol\,dm^{-3})$.

SAQ 1.1b
(Contd)

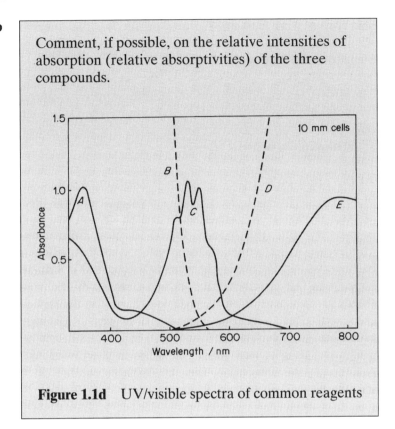

Comment, if possible, on the relative intensities of absorption (relative absorptivities) of the three compounds.

Figure 1.1d　UV/visible spectra of common reagents

Response

Solution	Spectrum
1	D
2	E
3	B
4	A
5	C

The normal reagent strength solutions of copper sulfate (blue), potassium dichromate (yellow) and potassium permanganate (deep purple) are so intensely absorbing that they show virtually complete absorption over certain regions of the visible spectrum (i.e. they absorb over 99.9% or produce absorbance values in excess of 3 when measured in 10 mm cells).

Therefore, $0.4 \, mol \, dm^{-3}$ copper sulfate solution (approximately 10% (wt/vol)) absorbs strongly above 640 nm (curve D) but when diluted to one tenth of the concentration the absorption maximum is seen to lie at 780 nm in the near IR! (curve E).

At the other end of the spectrum a solution of potassium dichromate of strength typical of that used in titrimetric analysis ($0.02 \, mol \, dm^{-3}$ or above) shows strong absorption in the visible region up to 500 nm (curve B) and this needs to be diluted approximately to one hundredth of that concentration before the absorption maximum can be observed at 350 nm in the near UV (curve A). Similar to the dilute copper sulfate solution this dilute potassium dichromate solution shows low absorption in the visible region and hence both are weakly coloured. The particular specification given for the solution of potassium dichromate ($100 \, mg \, dm^{-3}$) in dilute sulfuric acid ($0.005 \, mol \, dm^{-3}$) is that specified when the solution is to be used as a standard substance for checking the absorbance scale of a UV/visible spectrometer.

Solutions of potassium permanganate at concentrations used for titrimetric analysis (also typically $0.02 \, mol \, dm^{-3}$ or above) show virtually complete absorption across the whole visible spectrum with only narrow transmission bands at 420 and 760 nm (not shown), which produces the deep purple colour that is observed. Such strong solutions also need to be diluted to one hundredth of that concentration in order to give the measurable absorption maximum at 530 nm shown in curve C. The multiple-peak appearance of curve C is a unique characteristic of the permanganate ion absorption (see Chapter 5 of this present Unit).

From the above figure is it possible to read off approximate values of the absorbances of the three solutions at the wavelengths of maximum absorption and to calculate the molar absorptivities by using the Beer's law expression (see later in the main text). You could try these calculations to see if you can obtain answers close to the literature values noted below. Don't worry if you can't, as we will be covering this in a later section.

$$CuSO_4, 25\,dm^3\,mol^{-1}\,cm^{-1} \text{ at } 780\,nm$$

$$K_2Cr_2O_7, 250\,dm^3\,mol^{-1}\,cm^{-1} \text{ at } 351\,nm$$

$$KMnO_4, 2300\,dm^3\,mol^{-1}\,cm^{-1} \text{ at } 523\,nm$$

SAQ 1.2a	Carry out the following calculations:

(i) Obtain a value for the absorbance of a solution which only transmits 12% of the incident light.

(ii) Calculate the percentage of light transmitted for a solution with an absorbance value of 0.55.

(iii) Determine the value for the absorbance of a solution of an organic dye ($0.0007\,mol\,dm^{-3}$) in a cell with a 2 cm pathlength if its absorptivity is $650\,dm^3\,mol^{-1}\,cm^{-1}$.

Response

$A = \varepsilon c L$

In all cases it is a matter of using the numerical values for the appropriate terms of the following equations:

$$A = \log \frac{1}{T} = \log \frac{I_0}{I} = acl$$

(i)
$$A = \log \frac{100}{12} = \log 8.333$$

$$= 0.92$$

(ii) $$0.55 = \log \frac{100}{I}$$

$$\log 3.548 = \log \frac{100}{I}$$

$$I = \frac{100}{3.548}$$

$$I = 28\%$$

(iii) $$A = 650 \times 0.00070 \times 2$$

$$A = 0.91$$

If you have followed the steps in the development of the equation you should have found this fairly straightforward at this stage.

SAQ 1.2b   Calculate the concentration, in units of $mg\,dm^{-3}$, of a solution of each of the two compounds A and B.

Compound	M_r	ϵ $(dm^3\,mol^{-1}\,cm^{-1})$	Absorbance
A	250	1 000	0.10
B	250	100 000	0.10

What is the significance of the molar absorptivity in analysis?

Response

Using equation (1.12) $\epsilon = A/cl$

$$c = A/\epsilon l$$

Compound A: $c = 0.10/1000 \times 1$

$\qquad = 1 \times 10^{-4}\,\text{mol}\,\text{dm}^{-3}$

$\qquad = 250 \times 10^{-4}\,\text{g}\,\text{dm}^{-3}$

$c = 25.0\,\text{mg}\,\text{dm}^{-3}$

Compound B: $c = 0.10/100\,000 \times 1$

$\qquad = 1 \times 10^{-6}\,\text{mol}\,\text{dm}^{-3}$

$\qquad = 250 \times 10^{-6}\,\text{g}\,\text{dm}^{-3}$

$c = 25 \times 10^{-2}\,\text{mg}\,\text{dm}^{-3}$

Clearly, the greater the absorptivity, then the lower will be the concentration of analyte needed to give a measurable signal. You won't be surprised to learn that β-carotene, having a molar absorptivity similar in magnitude to compound B, is highly coloured and only small concentrations are needed to give the intense yellow-orange colour of carrots.

SAQ 1.3   Figure 1.3f below shows the optical components and layout of a typical recording spectrometer designed to operate over the wavelength range from 190 to 900 nm. Identify on the diagram the four principal components:

(i)   source;

(ii)   monochromator;

(iii)   sampling area;

(iv)   detector.

When the instrument is operated over certain wavelength ranges, filters are inserted into the optical path. Specify which of the filters, red or blue, is used:

(i)   at 780 nm;

(ii)   at 390 nm.

In each case, briefly explain the function of the filter used. What special precautions would you adopt to ensure the optimum instrument performance when measurements are being taken at 195 nm?

SAQ 1.3
(Contd)

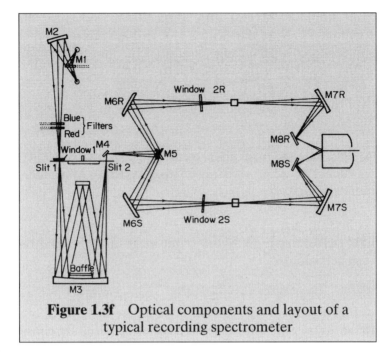

Figure 1.3f Optical components and layout of a typical recording spectrometer

Response

The annotated optical diagram of the spectrometer is shown below.

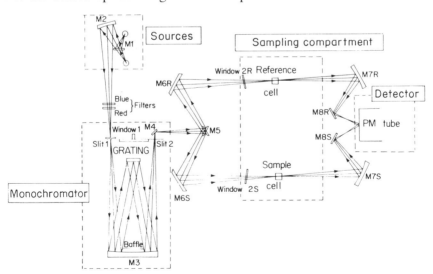

Figure 1.3g Identification of the principal components of a typical recording spectrometer

You should have had little difficulty in indicating the four main components. Note that with a UV/visible spectrometer the use of two sources to cover the range is normal. A tungsten or tungsten-halogen source is usual for the range 330 to 900 nm (and beyond), although the precise change-over point in the ultraviolet region can often be varied over the range from 325 to 385 nm. At lower wavelengths a deuterium lamp is normally used.

At 780 nm, the red filter would be used to prevent second-order wavelengths (e.g. 390 nm) from passing through the grating monochromator.

At 390 nm, the blue/violet filter would be used to improve the stray light performance by filtering out most of the visible range from the tungsten source.

When operating at 195 nm it is necessary to purge the instrument with nitrogen as the oxygen in the atmosphere absorbs at this wavelength. In addition, cells and solvents have to be carefully chosen in order to ensure adequate transmittance at this low wavelength. In the absence of a double monochromator, stray radiation effects will be significant at this wavelength (see later in this chapter).

SAQ 1.4	Indicate which of the following statements are *true* and which are *false*:
	(i) Quantitative colorimetric analysis requires the use of a light-measuring instrument.
	(ii) The eye is as good as an instrument for detecting colour changes.
	(iii) All colorimetric methods of analysis have been developed for trace levels.

SAQ 1.4
(Contd)

(iv) Certain biochemical methods of analysis can be designed to be highly selective, even though they all involve the measurement of change in the UV spectrum of nicotinamide–adenine dinucleotide (NAD).

(v) The use of UV/visible spectroscopy in the monitoring of the separation of mixtures by high performance liquid chromatography (HPLC) is limited to those components which show strong absorption above 220 nm.

Response

(i)   FALSE

(ii) TRUE

(iii) FALSE

(iv) TRUE

(v)   FALSE

(i) The visual comparator methods described in Section 1.2 use the eye as the device for comparing colour intensities, and hence for performing quantitative colorimetric analyses. However, such a colour comparison can also be made with the simplest of equipment, e.g. a pair of test tubes for holding coloured solutions.

(ii) This is true since the eye is particularly good at detecting small colour differences. Instruments capable of competing with the eye in this task are modern sophisticated recording spectrometers of high precision.

(iii) Although many colorimetric methods have been designed for the analysis of absorbing species at low levels of concentration (typically 10^{-2} to 10^{-5} mol dm^{-3}), the higher levels of this range could not be classified as trace levels. In addition, many spectroscopic methods of analysis involve significant dilutions for systems containing the components to be analysed at up to molar levels of concentration.

(iv) Biochemical analyses utilising NAD or its phosphate derivative, NADP, normally involve an enzymatic species which is specific for the component being analysed. The change in the UV spectrum occurs typically as a secondary but quantitative reduction/oxidation reaction of the NAD or NADP with the products of the enzyme conversion (an example is given in Chapter 3 of this present Unit).

(v) This is false since it is now common practice to label chemically, e.g. with UV absorbing or fluorescing groups, the components to be separated. In addition, the performance of UV monitors has been improved in the low-UV range to enhance the range of applications.

SAQ 1.5

> List two of the problems that might occur when calibration data do not obey Beer's law over the concentration range of interest.

Response

(i)   Drawing an accurate curve.

(ii) Interpolating accurately in the non-linear regions.

(iii) Uncertainty about the behaviour of sample solutions whose composition may be quite different from those of the standard solutions. It is possible, of course, for standard solutions to give a linear Beer's law relationship, and for the sample solutions to give a non-linear relationship due to the presence of non-analyte components arising from the sample matrix itself.

SAQ 2.1 | A biochemical enzymatic analysis is being carried out at 340 nm by spectroscopic measurements. Indicate which of the following would result in a large (L) and which would result in a small (S) effect on the measured absorbance.

(i) The sample becomes cloudy due to poor solubility. L/S

(ii) The sample is accidently placed in a glass cell instead of a silica cell. L/S

(iii) The sample cell is accidently contaminated with propanone. L/S

(iv) A tungsten source is used instead of a deuterium source. L/S

(v) The pH of the reaction system is not adjusted to the optimum value. L/S

Response

(i) L. Cloudiness results in a significant scatter of UV radiation, and hence a significant increase in apparent absorbance is expected.

(ii) S. This would probably have a small effect because 340 nm is significantly above the glass cut-off point, which occurs at about 310 nm. However, the effect when using plastic cells would be much greater due to their higher cut-off wavelengths.

(iii) L. Propanone would have a large effect since 340 nm is near the cut-off wavelength of 331 nm quoted in Table 2.1. In addition, an organic solvent such as propanone could have an influence on enzymatic reactions.

(iv) S. 340 nm is within the acceptable change-over range between the sources, although stray radiation may be more significant with the tungsten source (see Section 1.5).

(v) L. Although this topic has not been specifically mentioned your general knowledge of enzyme behaviour should be sufficient for you to appreciate the importance of pH control in this type of analysis (see Section 3.1), with optimum conditions being essential for correct analyses.

SAQ 2.2

Potassium thiocyanate and 1,10-phenanthroline have both been used as reagents for the determination of low concentrations of iron. Both have advantages and disadvantages for this application. Assign as many as possible of the following advantages and disadvantages to the two reagents.

Advantages

(i)   Complex formation requires only the addition of the reagent and some acid.

(ii) The reagent is cheap.

(iii) The complex is stable and relatively free from interferences.

(iv) The molar absorptivity is over $1000 \, m^2 \, mol^{-1}$.

(v)   It is applicable to iron in the Fe(III) state.

Disadvantages

(vi) The iron must be reduced to the Fe(II) state.

(vii) The complex is non-stoichiometric.

(viii) The molar absorptivity is too low to be chosen for water analysis (in the UK).

(ix) A control of pH is important.

(x)   The complex is sensitive to light, and is relatively unstable.

Response

Potassium thiocyanate   i, ii, v, vii, viii and x

1,10-Phenanthroline   iii, iv, vi, viii and ix

(see Section 2.2 for details).

Of these you may have been unable to assign (i) because you did not have a full set of data for the analytical procedure, and (iii) as no indication of possible interferences for 1,10-phenanthroline has been given.

SAQ 2.3a

Calculate the absorptivities of $KMnO_4$ using the following data:

A $KMnO_4$ solution at λ_{max} = 522 nm gave an absorbance of 1.236 in a 10 mm cell.

The Mn concentration is 30 mg dm^{-3} ($A_r(Mn)$ = 54.938).

(i) Molar absorptivity, ϵ_{max};

(ii) absorptivity, $E_{1\%}^{1\,cm}$.

Response

(i) The molar concentration of $KMnO_4$ = 30 × 10^{-3}/54.938 mol dm^{-3}
     (1 mole $KMnO_4$ contains 1 mole Mn)

$$= 546 \times 10^{-6} \, mol \, dm^{-3}$$

$$\epsilon_{max} = 1.236/(546 \times 10^{-6} \times 1)$$

$$= 2264 \, dm^3 \, mol^{-1} \, cm^{-1}$$

$$= 226 \, m^2 \, mol^{-1}$$

You will recall a previous calculation with slightly different data gave 230 m^2 mol^{-1}.

(ii) The % concentration is calculated as follows:

Mn concentration is $30 \times 10^{-3} \, \text{g dm}^{-3}$

$$= 3 \times 10^{-3} \, \text{g per 100 cm}^3 \, (\% \, (\text{wt/vol}))$$

$$\text{KMnO}_4 \text{ concentration} = 3 \times 10^{-3} \times 158.032/54.938$$

$$= 8.63 \times 10^{-3} \, \% \, (\text{wt/vol})$$

$$E_{1\%}^{1\,\text{cm}} = 1.236/(8.63 \times 10^{-3} \times 1)$$

$$= 143 \, \text{cm}^2 \, \text{g}^{-1}$$

SAQ 2.3b

A solution of potassium permanganate with a manganese concentration of $3.4 \, \text{mg dm}^{-3}$ transmits 23% at 522 nm and 57.5% at 480 nm. Calculate the effect of a $+1\%$ transmittance error on the absorbance at these two wavelengths. Indicate, giving reasons, which wavelength it is best to use for the analysis of permanganate solutions (522 nm is the λ_{max} for KMnO_4; 480 nm is on the side of the adsorption band).

Response

At $\lambda = 522 \, \text{nm}$,

Absorbance value:
$$A = \log 100/23 = 0.638$$
$$A = \log 100/24 = 0.620$$
$$\Delta A = 0.018$$

At $\lambda = 480 \, \text{nm}$,

$$A = \log 100/57.5 = 0.240$$
$$A = \log 100/58.5 = 0.233$$
$$\Delta A = 0.007$$

The % absorbance errors are, respectively, as follows:

$$0.018/0.638 = 2.8\%$$

$$0.007/0.240 = 2.9\%$$

Therefore, the percentage errors are comparable, under the above conditions.

Measurements are best taken at 522 nm for the following reasons:

(i) The errors due to wavelength shifts are minimised.

(ii) The calibration graphs have greater slopes, hence the concentration read-out errors are minimised.

(iii) The calibration graphs show less scatter at the λ_{max} values.

SAQ 2.4	A 10.0 cm³ aliquot of an aqueous solution of quinine was diluted to 25 cm³ and found to have an absorbance of 0.217 at 348 nm when measured in a 1.00 cm pathlength cell. A second 10.0 cm³ aliquot was mixed with 5.00 cm³ of a solution containing 27.3 parts per million (ppm) of quinine. After dilution to 25.0 cm³ this solution had an absorbance of 0.474 when measured in the same 1.00 cm pathlength cell. Calculate the amount of quinine, in ppm, in the original aqueous solution.

Response

We can use equation (2.7) as follows:

$$C_x = \frac{A_1 C_s V_s}{(A_2 - A_1)V_x}$$

where:

$$A_1 = 0.217$$

$$A_2 = 0.474$$

$$C_S = 27.3\,\text{ppm}$$

$$V_S = 5.00\,\text{cm}^3$$

$$V_x = 10.0\,\text{cm}^3$$

Substituting these values into equation (2.7) we have:

$$C_x = \frac{0.217 \times 27.3 \times 5.00}{(0.474 - 0.217) \times 10.0}$$

$$= 11.5\,\text{ppm}$$

SAQ 3.1a

In many colorimetric analyses it is sufficient to correct the sample absorbance (S) for the blank reading (B), and to read off the component concentration from the Beer's law calibration graph. This is implied in the above procedure for iron in water when the apparent absorbance is given by:

$$R = S - B$$

Indicate if when using the above simple procedure the factors below would result in a high (H), low (L) or correct (C) value for the iron content of the water being analysed. If you have insufficient information to make a judgement use the code (I).

(i)   The temperature of the sample dropped to 15°C before measurement.

(ii)   The instrument was found to have a 1 nm calibration error.

(iii) The sulfuric acid reagent was found to contain $1 \, mg \, dm^{-3}$ iron.

(iv)   The deionised water used for the blank was found to contain $0.1 \, mg \, dm^{-3}$ iron.

(v)   Only 1/10 of the quantity of hydroxylammonium chloride reagent was added ($0.2 \, cm^3$ instead of $2.0 \, cm^3$).

(vi)   The final solution for measurement looked slightly cloudy.

Response

(i)   C. The TPTZ method for iron appears to be relatively temperature insensitive since the procedure merely requires the temperature to be adjusted to somewhere between 15 and 30°C. For certain procedures a low temperature may reduce the rate of achieving a stable reading.

(ii) I(C). Coloured complexes usually have fairly broad absorption maxima and therefore an error of 1 nm is unlikely to affect the measured absorbance. However, to be sure we would need to have information about the shape of the absorption curve in the vicinity of λ_{max}.

(iii) C. A small amount of iron in the reagents would be compensated for by the blank reading.

(iv) L. In this case, iron in the water used in the blank would result in a low value of the iron and this is allowed for in Section 9.7 and 9.8 of the HMSO procedure and by the quantity C_w in the calculation of the final result in 9.11.

(v) L(I). Small variations in the amounts of the complexing reagent are unlikely to affect the results as excess reagent is normally added. However, adding only 10% of the required amount is likely to be insufficient to complex all of the iron in the water. This is confirmed by consultation of the original report on the development of the analytical method in W. K. Dougan and A. L. Wilson, *The Determination of Iron in Water*, The Water Research Association, Marlow, 1972.

(vi) H. Cloudiness in the final solution probably indicates turbidity, and hence light scattering, thus leading to a higher than expected absorbance of the analytical light beam. The procedure 9.6 allows for the correction S_1 for scattering and gives:

$$R = S - B - S_1$$

If the scattering is due to suspended organic matter this should be eliminated by a wet-oxidation pretreatment process which is detailed in Section 8 of the HMSO publication (not given here).

SAQ 3.1b | The success of the enzymatic determination of glucose by the method described in Section 3.1.2 is implicit in the statement that 'NADPH is stoichiometric with the amount of glucose and is determined by means of its absorption at 334, 340 or 365 nm'. Why do you think three wavelengths are specified and what are the implications in terms of the instrumental precision which is achievable?

Response

Three wavelengths are specified in order to allow the use of instruments which employ filtered mercury light sources as well as those which employ the normal tungsten light source with a variable wavelength monochromator. With the latter the wavelength would be set for the λ_{max} of NADPH, i.e. 340 nm. With such instruments we normally avoid using the sloping side of an absorption band which is subject to wavelength setting and bandwidth errors (Section 1.5). A mercury line source does not suffer from wavelength errors (only focusing errors) but can have some other drawbacks in terms of adaptability to other types of analysis.

The method of calculation gives values of ϵ appropriate to the three wavelengths, but these ϵ-values may not be fully reproducible on different UV instruments!

SAQ 3.1c | The molar absorptivity (ϵ_{max}) for the ternary complex of iron(II), 4-chloro-2-nitrosophenol and Rhodamine B is quoted as $9.0 \times 10^3\,m^2\,mol^{-1}$ at 558 nm. This is only a factor of four times (4×) greater than that obtained with the iron(II) tripyridyltriazine (TPTZ) complex, yet the typical iron concentration levels measured are at least ten times lower (50 μg dm^{-3} against 500 μg dm^{-3}). How is this achieved?

Response

The principal reason is that the complex is extracted from 20 cm³ of aqueous solution into 5 cm³ of toluene prior to the measurements being carried out. However, this further fourfold improvement is eliminated by the choice of a 10 mm pathlength for the ternary complex in the toluene, compared to a 40 mm pathlength for the TPTZ method.

The minimum detectable level quoted for the TPTZ method is 3 to 15 $\mu g\,dm^{-3}$ Fe(II), corresponding to an absorbance of only 0.004 to 0.02. The lowest absorbance quoted in the *Analyst* report (not given here) is 0.042 ± 0.001, which corresponds to 0.13 ± 0.003 $\mu g\,dm^{-3}$. Such precision in absorbance measurements suggests that a modern instrument of high precision was used in the latter investigation.

SAQ 3.3

A certain dye A has the same absorption spectrum in aqueous solution containing either 10% pyridine or 10% ethanol. Dye B behaves in the same way. A mixture of the two dyes, however, does not give the same absorption spectrum in the two solvents. Determine from the data below the composition of the mixture, showing how the more satisfactory data are selected.

System	Absorbance		
	450 nm	550 nm	650 nm
Dye A ($1\,g\,dm^{-3}$)	1.75	0.68	0.21
Dye B ($1\,g\,dm^{-3}$)	0.11	0.23	0.33
Mixture in ethanol–water	2.07	1.75	1.30
Mixture in pyridine–water	2.08	1.37	1.20

(Figures given in the table are absorbances measured in a 10 mm cell at the wavelengths indicated.)

SAQ 3.3 (Contd)	To help you make a start, treat the problem as one involving a binary mixture, choose the two most appropriate wavelengths for the analysis, and then check that the result fits at the third wavelength. You will have to decide which solvent system to choose for the calculation — only one solvent gives satisfactory results.

Response

Treat the problem as a binary analysis at 450 and 650 nm and check that the result fits at 550 nm.

As it is the less polar solvent, pyridine is likely to cause less interaction than would ethanol. Therefore, we should set up simultaneous equations as follows:

$$\text{at } 450 \text{ nm}, \quad 2.08 = 1.75\, C_A + 0.11\, C_B$$

and

$$\text{at } 650 \text{ nm}, \quad 1.20 = 0.21\, C_A + 0.33\, C_B$$

$$\text{Solving gives } C_A = 1\, \text{g dm}^{-3} \text{ and } C_B = 3\, \text{g dm}^{-3}.$$

Checking at 550 nm, $0.68 + 3 \times 0.23 = 1.37$, as observed.

As the observed and calculated values agree for 550 nm, additive behaviour clearly applies in aqueous pyridine.

For ethanol/water mixtures the same procedure gives the following:

$$2.07 = 1.75\, C_A + 0.11\, C_B$$

$$1.30 = 0.21\, C_A + 0.33\, C_B$$

Solving in this case gives:

$$C_A = 0.98\,\text{g dm}^{-3} \text{ and } C_B = 3.28\,\text{g dm}^{-3}$$

Checking at 550 nm gives:

$$0.68 \times 0.98 + 3.28 \times 0.23 = 1.42 \neq 1.75$$

The observed and calculated values do not agree in this case and therefore the two dyes show non-additive behaviour in aqueous ethanol.

SAQ 4.1   Standard solutions of potassium permanganate were prepared with the following concentrations:

Solution	A	B	C	D
Concentration ($mg\,dm^{-3}$)	10.0	20.0	30.0	40.0

The second-order derivative spectra of these standards are shown in Figure 4.1l.

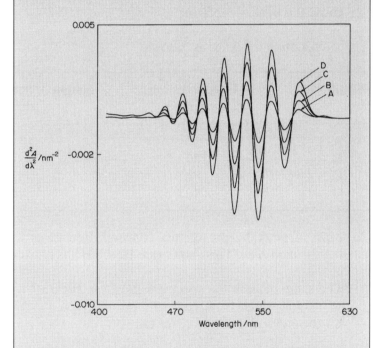

Figure 4.1l   Second-order derivative spectra of potassium permanganate standard solutions

$\rightarrow$

SAQ 4.1 (Contd)	(i) Show that these standards obey Beer's law. (ii) Use the second-order derivative spectrum of $KMnO_4$ shown in Figure 4.1g to calculate its concentration.

Response

(i) You need to measure the height (in mm) of one of the peaks. I chose the peak at ~ 550 nm.

The values obtained were as follows:

Solution	Concentration $(mg\,dm^{-3})$	Height of peak at 550 nm (mm)
A	10	13
B	20	27
C	30	41
D	40	56

A graph of peak height against concentration gives a straight line passing through zero. This shows that Beer's law is obeyed.

(ii) The same peak in the spectrum shown in Figure 4.1g has a height of 35 mm, which from your calibration graph should give a value for the concentration of 25 mg dm^{-3}.

SAQ 5.2   Figure 5.2 is an absorption spectrum of potassium permanganate in which the four major peaks have been numbered. Values for λ_{max} and ϵ_{max} for the strongest peak are given below. Measurements were made using a 1 cm cell.

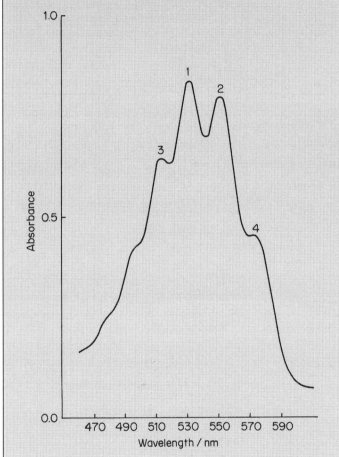

Figure 5.2   Absorption spectrum of potassium permanganate

→

SAQ 5.2
(Contd)

Complete the following table to show the λ_{max} and ϵ_{max} values of the three major bands.

Band	λ_{max} (nm)	ϵ_{max} (m² mol⁻¹)
1	533	1.01×10^3
2		
3		
4		

Response

The three major peaks for the potassium permanganate curve are easily obtained from the wavelength scale. The value for ϵ_{max} in each case is calculated from the following equation:

$$\epsilon = \frac{A}{cl}$$

The value for c required in the three calculations may be obtained from the band 1 values, with $l = 0.01$ m.

$$c = \frac{A}{\epsilon l}$$

$$= \frac{0.83}{1.01 \times 10^3 \times 0.01}$$

$$= 8.2 \times 10^{-4}\,\text{mol}\,\text{dm}^{-3}$$

The final results should be as follows:

Band	λ_{max} (nm)	ϵ_{max} (m² mol⁻¹)
1	533	1.01×10^3
2	552	0.96×10^3
3	514	0.78×10^3
4	574	0.54×10^3

SAQ 5.4
Assign from list 2 below, the appropriate values of λ_{max} and ϵ_{max} for each system given in list 1. Comment on the shifts observed relative to benzene (λ_{max} = 255 nm; ϵ_{max} = 23 m^2 mol^{-1}) in terms of the electronic properties of the auxochromes.

List 1 Absorbing systems:

A aqueous solution of aniline;

B solution A with 0.01 M HCl added;

C aqueous solution of phenol;

D solution C with 0.01 M NaOH added;

List 2 Absorption characteristics:

System	λ_{max} (nm)	ϵ_{max} (m^{2}mol^{-1})
E	255	20
F	270	140
G	280	130
H	290	230

Response

List 1–list 2 combinations (see below for comments on the alternatives listed):

A–F(G)

B–E

C–G(F)

D–H

The *chromophore* is the benzene ring in each case, while the *auxochromes* are NH$_2$ (A), NH$_3^+$ (B), OH (C) and O$^-$ (D).

The shifts relative to benzene (λ_{max} = 255 nm) depend on the electron-donating properties of the auxochromes, which are in the following order:

$$NH_3^+ \ll OH < NH_2 \ll O^-$$

This order is essentially the order which can be deduced from a knowledge of the basic chemistry of these groups in aromatic substitution. You might not have appreciated that the lone pair of electrons on the NH_2 group produces a greater red (bathochromic) shift than that on the OH group, thus giving the alternatives mentioned above. You should have been aware that the addition of acid to form NH_3^+ effectively removes the nitrogen lone pair, while the formulation of RO^- is generally associated with significant bathochromic shifts.

SAQ 5.5

A compound with the following structure:

is classed as (i) an alkene if X is CH_2, and (ii) an unsaturated ketone if X is O.

For the purpose of applying the Woodward rules to predict the λ_{max} of the UV spectrum in alcohol solution we need to be able to characterise certain aspects of the structural contributions.

(A) In this context, indicate which of the following statements are *true* and which are *false*.

(i) When X is CH_2 we have a triene and the diene rules do not apply.

SAQ 5.5
(Contd)

(ii) The compound has only *two* exocyclic double bonds.

(iii) The compound has two methyl groups, both of which contribute +5 nm to the absorption.

(iv) When X is an oxygen atom (O) the double bonds of the unsaturation are in the α, β and γ, δ positions.

(B) By using the appropriate rules, show that the predicted λ_{max} values in EtOH are as follows:

(i) 269 nm for the alkene structure;

(ii) 277 nm for the ketone structure.

Response

(A)

(i) FALSE; although this *is* a triene the diene rules *still* apply.

(ii) FALSE; there is only *one* exocyclic double bond, i.e. the central one of the triene.

(iii) FALSE; only *one* methyl group contributes to the system, although there are also *three* ring residues to take into consideration.

(iv) TRUE; the basic system is as follows:

$$\underset{H}{\overset{\delta}{\diagdown}}C = \underset{}{\overset{\gamma}{C}} - \underset{\beta}{\overset{|}{C}} = \underset{H}{\overset{\alpha}{C}} - \overset{|}{C} = O$$

(B) The wavelengths are obtained as follows:

Component	(i)	(ii)
Acyclic diene	214	–
α,β-Unsaturated carbonyl unit	–	222
Extended conjugation	30	30
Exocyclic double bond	5	5
Methyl group	5	5
Ring residues (3)	15	15
Total	269 nm	277 nm

Units of Measurement

For historical reasons a number of different units of measurement have evolved to express a quantity of the same thing. In the 1960s, many international scientific bodies recommended the standardisation of names and symbols and the adoption universally of a coherent set of units — the SI units (Système Internationale d'Unités) — based on the definition of seven basic units: metre (m), kilogram (kg), second (s), ampere (A), kelvin (K), mole (mol), and candela (cd).

The earlier literature references and some of the older text books naturally use the older units. Even now many practising scientists have not adopted SI units as their working units. It is, therefore, necessary to know of the older units and to be able to interconvert these with the SI units.

In this series of texts SI units are used as standard practice. However, in areas of activity where their use has not become general practice, for example biologically based laboratories, the earlier defined units are used. This is explained in the study guide to each unit.

Table 1 shows some symbols and abbreviations commonly used in analytical chemistry, while Table 2 shows some of the alternative methods for expressing the values of physical quantities and their relationship to the values in SI units. In addition, Table 3 lists prefixes for SI units and Table 4 shows the recommended values of a selection of physical constants.

Further details and definitions of other units may be found in I. Mills, T. Cvitaš, K. Homann, N. Kallay and K. Kuchitsu, *Quantities, Units and Symbols in Physical Chemistry*, 2nd Edn, Blackwell Science, 1993.

Table 1   Symbols and abbreviations commonly used in analytical
chemistry

Å	angstrom
$A_r(X)$	relative atomic mass of X
A	ampere
E or U	energy
G	Gibbs free energy (function)
H	enthalpy
I (or i)	electric current
J	joule
K	kelvin $(=273.15 + t(°C))$
K	equilibrium constant (with subscripts p, c, etc.)
K_a, K_b	acid and base ionisation constants
$M_r(X)$	relative molecular mass of X
N	newton (SI unit of force)
P	total pressure
s	standard deviation
T	temperature (K)
V	volume
V	volt (J A^{-1} s^{-1})
$a, a(A)$	activity, activity of A
c	concentration (mol dm^{-3})
e	electron
g	gram
s	second
t	temperature (°C)
b.p.	boiling point
f.p.	freezing point
m.p.	melting point
~	approximately equal to
<	less than
>	greater than
e, $\exp(x)$	exponential of x
$\ln x$	natural logarithm of x; $\ln x = 2.303 \log x$
$\log x$	common logarithm of x to base 10

Table 2 Summary of alternative methods of expressing physical quantities

---

(1) Mass (SI unit: kg)

$$g = 10^{-3}\,\text{kg}$$
$$mg = 10^{-3}\,\text{g} = 10^{-6}\,\text{kg}$$
$$\mu g = 10^{-6}\,\text{g} = 10^{-9}\,\text{kg}$$

(2) Length (SI unit: m)

$$cm = 10^{-2}\,\text{m}$$
$$\text{\AA} = 10^{-10}\,\text{m}$$
$$nm = 10^{-9}\,\text{m} = 10\,\text{\AA}$$
$$pm = 10^{-12}\,\text{m} = 10^{-2}\,\text{\AA}$$

(3) Volume (SI unit: m³)

$$l = \text{dm}^3 = 10^{-3}\,\text{m}^3$$
$$ml = \text{cm}^3 = 10^{-6}\,\text{m}^3$$
$$\mu l = 10^{-3}\,\text{cm}^3$$

(4) Concentration (SI unit: mol m⁻³)

$$M = \text{mol}\,l^{-1} = \text{mol}\,\text{dm}^{-3} = 10^3\,\text{mol}\,\text{m}^{-3}$$
$$mg\,l^{-1} = \mu g\,\text{cm}^{-3} = \text{ppm} = 10^{-3}\,\text{g}\,\text{dm}^{-3}$$
$$\mu g\,g^{-1} = \text{ppm} = 10^{-6}\,\text{g}\,g^{-1}$$
$$ng\,\text{cm}^{-3} = \text{ppb} = 10^{-6}\,\text{g}\,\text{dm}^{-3}$$
$$pg\,g^{-1} = \text{ppt} = 10^{-12}\,\text{g}\,g^{-1}$$
$$mg\% = 10^{-2}\,\text{g}\,\text{dm}^{-3}$$
$$\mu g\% = 10^{-5}\,\text{g}\,\text{dm}^{-3}$$

(5) Pressure (SI unit: $N\,m^{-2} = kg\,m^{-1}\,s^{-2}$)

$$Pa = N\,m^{-2}$$
$$atm = 101\,325\,N\,m^{-2}$$
$$bar = 10^5\,N\,m^{-2}$$
$$torr = mmHg = 133.322\,N\,m^{-2}$$

(6) Energy (SI unit: $J = kg\,m^2\,s^{-2}$)

$$cal = 4.184\,J$$
$$erg = 10^{-7}\,J$$
$$eV = 1.602 \times 10^{-19}\,J$$

Table 3   Prefixes for SI units

Fraction	Prefix	Symbol
10^{-1}	deci	d
10^{-2}	centi	c
10^{-3}	milli	m
10^{-6}	micro	μ
10^{-9}	nano	n
10^{-12}	pico	p
10^{-15}	femto	f
10^{-18}	atto	a

Multiple	Prefix	Symbol
10	deca	da
10^{2}	hecto	h
10^{3}	kilo	k
10^{6}	mega	M
10^{9}	giga	G
10^{12}	tera	T
10^{15}	peta	P
10^{18}	exa	E

Table 4 Recommended values of physical constants

Constant	Symbol	Value
acceleration due to gravity	g	$9.81\,\mathrm{m\,s^{-2}}$
Avogadro constant	N_A	$6.022\,14 \times 10^{23}\,\mathrm{mol^{-1}}$
Boltzmann constant	k	$1.380\,66 \times 10^{-23}\,\mathrm{J\,K^{-1}}$
charge-to-mass ratio	e/m	$1.758\,796 \times 10^{11}\,\mathrm{C\,kg^{-1}}$
electronic charge	e	$1.602\,18 \times 10^{-19}\,\mathrm{C}$
Faraday constant	F	$9.648\,46 \times 10^{4}\,\mathrm{C\,mol^{-1}}$
gas constant	R	$8.314\,\mathrm{J\,K^{-1}\,mol^{-1}}$
ice-point temperature	T_{ice}	$273.150\,\mathrm{K}$, exactly
molar volume of ideal gas (stp)	V_m	$2.241\,38 \times 10^{-2}\,\mathrm{m^3\,mol^{-1}}$
permittivity of a vacuum	ε_0	$8.854\,188 \times 10^{-12}$ $\mathrm{kg^{-1}\,m^{-3}\,s^4\,A^2}$ ($\mathrm{F\,m^{-1}}$)
Planck constant	h	$6.626\,08 \times 10^{-34}\,\mathrm{J\,s}$
standard atmosphere (pressure)	p	$101\,325\,\mathrm{N\,m^{-2}}$, exactly
atomic mass constant	m_u	$1.660\,54 \times 10^{-27}\,\mathrm{kg}$
speed of light in a vacuum	c	$2.997\,925 \times 10^{8}\,\mathrm{m\,s^{-1}}$

Index